AF807043

Some Critical Questions in Biological Physics

A guided tour around the bugbears

The front page illustration (following the picture frames left to right and then top to bottom) includes: a cellular membrane, a *Spirochete* bacterium (that causes Leptospirosis), a model polyelectrolyte chain (e.g. DNA), motor proteins inside a cell that move an endosome, an electrical spike that travels in a neuronal cell in the brain, starch in a dry uncooked potato, a chromosome, a slug with a mucus trail, a quantum mechanically entangled compass in the brain of a bird (Schrödinger's duck) and a polar bear.

Some Critical Questions in Biological Physics

A guided tour around the bugbears

Thomas Andrew Waigh

Biological Physics, School of Physics and Astronomy, Schuster Building, University of Manchester, UK

IOP Publishing, Bristol, UK

© IOP Publishing Ltd 2017

All rights reserved. No part of this publication may be reproduced, stored in a retrieval system or transmitted in any form or by any means, electronic, mechanical, photocopying, recording or otherwise, without the prior permission of the publisher, or as expressly permitted by law or under terms agreed with the appropriate rights organization. Multiple copying is permitted in accordance with the terms of licences issued by the Copyright Licensing Agency, the Copyright Clearance Centre and other reproduction rights organisations.

Permission to make use of IOP Publishing content other than as set out above may be sought at permissions@iop.org.

Thomas Andrew Waigh has asserted his right to be identified as the author of this work in accordance with sections 77 and 78 of the Copyright, Designs and Patents Act 1988.

ISBN 978-0-7503-1374-2 (ebook)
ISBN 978-0-7503-1375-9 (print)
ISBN 978-0-7503-1376-6 (mobi)

DOI 10.1088/978-0-7503-1374-2

Version: 20170901

IOP Expanding Physics
ISSN 2053-2563 (online)
ISSN 2054-7315 (print)

British Library Cataloguing-in-Publication Data: A catalogue record for this book is available from the British Library.

Published by IOP Publishing, wholly owned by The Institute of Physics, London

IOP Publishing, Temple Circus, Temple Way, Bristol, BS1 6HG, UK

US Office: IOP Publishing, Inc., 190 North Independence Mall West, Suite 601, Philadelphia, PA 19106, USA

For my daughter, Emily. I hope you will always keep your sense of wonder.

Contents

Preface

Biological physics is a relatively recent addition to the standard roster of subject areas taught in physics courses and it can be contrasted against more traditional areas such as atomic physics, high energy physics, optical physics, astrophysics or solid state physics. Historically biophysics has been extremely successful, providing crucial contributions to our understanding of the world around us, but in most cases physicists who did biology in the twentieth century were encouraged to leave physics departments to become dedicated life scientists. Perhaps most famously this occurred to the early crystallographers, whose work led to the establishment of whole new fields in the life sciences, which prompted the creation of purpose built laboratories for molecular biology. However, from a modern perspective, it is thought that keeping some physicists who study biology in actual physics departments will improve the transfer of ideas between the specialities. Over specialization can lead to improved efficiency in the short term, but can lead to inflexibility and short sightedness on longer time scales.

It is worthwhile to be more specific considering the words used to define the different subject areas that will be discussed. In terms of etymology the word physics means *nature* (or origin), whereas physiology (the old-fashioned discipline associated with biological physics) is almost the same word; it means the *study of nature* (or origin). Chemistry, in contrast, is defined as the study of matter.

The branding *biological physic*s has been used to separate the discipline from medical physics (physicists collaborating with medics in a clinical setting), biophysics (often adopted by life scientists who use physical tools), physiology (e.g. medics trying to understand the physics of humans) and mathematical biology (mathematicians collaborating with biologists). In reality, these divisions are fairly artificial, and all the disciplines provide valuable contributions and perform important functions.

In common with other sub-disciplines within physics, biological physics has a series of mysteries associated with it that motivate research scientists to turn up to work in the mornings in a bid to solve them. The description of some of the key mysteries in biological physics is the goal of this book.

Currently I work as an academic in a physics department at the University of Manchester in the UK. My colleagues take the business of pure physics very seriously. This means that we biological physicists are often placed in the ludicrous situation that we need to argue the case for biophysics among other purer physics disciplines; particularly high energy physics and astrophysics here in Manchester. Appealing to the impact of the research on mankind's health and wellbeing is normally a sufficient *raison d'être* for the rest of the human race, but the physicists need to be persuaded that the field actually contains some important physics, and it provides an interesting perspective to defend the existence of biological physics research under this line of questioning, treating it as a purely academic pursuit. Pure physics is often seen as a delicate, precious flower that must be protected at all costs, whereas applied physics (e.g. biological physics) is a vulgar pursuit performed by money grabbing barbarians, and it is thought to be big enough and ugly enough to fend for itself. Even worse, applied physics could distract good scientists from more important pure research topics, such as the existence of dark matter or the exotic flavours of neutrinos.

I was a little taken aback when I first needed to make the case for biological physics. This was then followed by the fall-back position of vaguely trying to communicate my interest in the subject and arguing the case for medical science. However, was I really advocating that we retrain physicists to become some fancy new type of technical nurses? Now my argument has changed. Active areas of physics research tend to be defined by their unsolved problems, since as an area becomes more mature and better determined it inevitably becomes engineering research not physics. An old adage is that physics is when something works once, whereas engineering is when you need it to happen all the time. We thus should ask the question of what has not yet been done; not even once? So, my new argument is that biological physics can claim the most important unsolved problem in science, the inner workings of the brain, which includes such phenomena as intelligence and consciousness. In addition, there are a whole host of other important problems in quantitative biology (hopefully slightly easier) that stand up well to comparison with anything the domain of pure physics has on offer, e.g. the kinetics of gene expression or the electrical waves in heart tissue. In this book I will therefore try to describe some of these unsolved problems in biological physics, which in turn will hopefully help to better define the field.

Again, trying to respond to comments from pure physics colleagues questioning the existence of biophysics, it is wrong that everyday science (the bits that are left when you cut off all the esoteric material) is almost complete, since there are huge gaps in our knowledge in most fields. Although we have some good theories for elementary particles, our understanding of how they interact in large numbers contains significant gaps, and the prospects of using elementary particle theories to make useful predictions in a biological problem, such as the folding pattern of chromatin in the nucleus of a cell, are practically non-existent. We need pragmatic physical models that predict phenomena that directly impact on our everyday lives, not additional intractable many-body calculations.

I hope the unsolved problems described in this book will be answered in the near future (within my life time) and the expectation is that their solutions will be of great benefit to mankind. Furthermore, biological physics is a good field for young researchers to learn the scientific method, the day-to-day practicalities of how to do science. This can be favourably compared with high energy physics or astrophysics that are often completely hands-off. Students wait for the new data sets to be sent from the CERN hadron collider or the Hubble telescope, without the ability to directly interact with their experiments in any tangible manner. I should stress that I find many of the traditional areas of pure physics, such as high energy physics and astrophysics, to be fascinating subjects and I expect they will have long rosy futures. However, here I make the case for biological physics research and one convenient way to do that is to contrast it against other fields.

Biological physics tends to be an interdisciplinary area, which gives it a distinct character when compared with other more traditional subdivisions of physics, e.g. a mixture of physics, mathematics, engineering, chemistry, medicine and biology can all play an important role. It is, however, not unique in this respect, since most traditional areas of physics become multi-disciplinary when they encounter real world problems. For example, consider atomic physics for new light sources. What is the chemistry of the lasing material? How can the light output be optimized mathematically? How do you engineer the ideal low noise power supply? How does the spectral balance of the light affect the physiology of human vision? This was Douglas Thomas Waigh's (my grandfather) job. He was an atomic physicist who worked on fluorescent lighting at the Thorn Ltd company in London, later called Thorn EMI. 'A jack of all trades' is how he described himself. Another example of a real world problem is nuclear physics for clean energy. What is the chemistry of the radionuclides? What is the best engineering design for the plant? What are the material properties of the fuel rods? What is the ecological impact of mutated microorganisms? In fact, bacteria constantly mutate at a high rate under natural conditions, so mutated bugs are probably not a serious issue. *Radiodurans*, an extremely radiation hardy bacterium (it can withstand a radiation dose five hundred times as much as a human), was first isolated in γ-ray sterilisation experiments with food using a radioactive source, although it is not clear how the bacterium evolved to be so radiation resistant in the first place.

All real world applications of physics are thus invariably interdisciplinary. There is a mixed martial arts analogy to interdisciplinary science, which I apologise for in advance to people with delicate sensibilities. In a fight with a problem in the real world, the person with the broadest range of skills is often in the best position to finish off the problem; there is no umpire that disqualifies participants if someone happens to bring an exotic type of mathematics, physics or chemistry to the table. Researchers are not placed on a level playing field, with penalties to insure it is a fair fight if some of them have broader non-standard academic backgrounds than others. The person with the most diverse skill set is often best placed to solve problems within the broadest remit and such problems are common in biological physics. Of course, this is a loose analogy that should not be taken too far, since the scientific equivalent of cage fighting does occasionally happen, but it is certainly not recommended. Coincidentally, the physical concept of cages is well established in the fields of both colloidal hydrodynamics and liquid state physics, which have many applications in biology and have reputations for both vociferously held opinions and combative researchers.

I believe the biological physics problems described in the book are of critical importance to the field. However, to discuss them with any authority, the biological physics problems chosen are predominantly a compilation of some personal bugbears that myself and members of our community have had to live with for a number of years. Thus a selection of unsolved problems will be presented, with different levels of difficulty, that range from practically unsolvable questions, such as the many-body problem with a long-range interaction, or the Navier–Stokes equations in an arbitrary geometry, to some much more modest targets to build up a researcher's confidence. Hopefully the simpler problems will appeal to people who are new to the world of scientific research or just want to sample some low hanging fruit in a new field in the shortest possible time. Indeed, some problems in zoology and food science can even be solved with household measurement apparatus (e.g. a microwave and a food blender), that highlight the approachability of certain forms of biophysics research, which can be favourably contrasted with some mega-scale science projects that have attracted large (billion dollar) investments, such as the human genome project, that often dominate the science headlines.

We will mostly focus on known unknowns in the biophysics questions posed, as opposed to unknown knowns (when important data is lost, e.g. a hard drive is corrupted on a computer or the last living expert finally pops their clogs) or the deeply mysterious and slightly oxymoronic unknown unknowns (huge areas of knowledge that we do not yet have the slightest inkling of our huge levels of ignorance). Amongst the vast edifice of current knowledge, we can lose sight of many important unknowns, hidden among a huge selection of run-of-the-mill knowns. Clearly an immense selection of important problems will slip through the gaps in the following discussion and the emphasis will be on my own personal bugbears, as opposed to problems chosen using more quantitative criteria, e.g. diseases in the world ranked by mortality rates. These lists of diseases tend to be well covered in the traditional medical and biological research literatures.

A little should be said about objectivity in science. Everyone of course likes to think they are completely objective in the work that they perform. However, earnest examination of scientific rigour sometimes goes out of the window in modern science. A peculiar modern phenomenon is that of counter intelligence in research. This tends to involve Milo Minderbinder characters that try to sell the research equivalent of chocolate cotton to the masses. Milo is a fictional character from Joseph Heller's *Catch 22* and was written as an ethically unsound war profiteer, whose activities defied all logic other than their ability to make him lots of money, e.g. selling chocolate cotton that no-one likes and no-one can eat. The product of Milo's research cannot possibly be faulty or misguided because it was given lots of funding, has the backing of eminent professors or (even worse) is the product of an eminent professor's life's work. Milo sounds like an amusing harmless character, but problems can occur when he starts to take his counter intelligence skills seriously, e.g. degrading the supply lines of his competitors (such as abusing the peer review system on grant applications), engaging the perceived enemies of chocolate cotton in a propaganda war or just using his immense knowledge of reactionary politics to lay waste to the opposition. Everybody else in fact works on chocolate cotton, although his research does involve substantial quantities of cellulosic fabrics and Cadbury's flakes. The victim can thus easily become Milo's axolotl, if they complain, as he reverses the tables on them (another literary reference, this time to a short story by Julio Cortazar). Clearly the situation can quickly start to get out of hand. The bottom line should be the quality of the science and it is hoped that the following discussion predominantly circumvents the interference of too many Milos.

My father is an organic chemist, or more specifically, a medicinal chemist who worked on natural products; someone who tries to make new drugs that are derived from naturally occurring molecules found in plants. My father's main claim to fame is as the co-inventor of the drug Atracurium, which is now widely used in hospitals as a general anaesthetic. Atracurium replaced Curare, which was originally used by Native Americans on poisoned arrows, but the dosage of Curare is hard to control in a clinical setting, unlike Atracurium, hence its modern usage. A glamorous aspect to my dad's work involved travel to different parts of Africa to collect botanical samples and talk to witch doctors (one man's witch doctor is another man's ultrahigh technology medicinal chemist, so this is a matter of perspective), but his working life predominantly was that of a synthetic chemist, who put plant extracts down chromatography columns and into NMR machines, and then tried to figure out how to mass produce the useful ones the team were lucky enough to identify. The variety of molecules found in the plants was huge and he was consequently left with a vast list of unsolved problems when he retired a few years ago, his own personal range of bugbears. Hydrogen bonding patterns in molecules were particularly high on his list, along with minority side effects of perfectly good drugs that could have been rectified by pretesting, i.e. by using what is now called personalized medicine. To avoid this situation myself, here I would quite selfishly like to catalogue some of my own bugbears before I too am put out to pasture in twenty five years' time, and I hope the additional publicity might help a few of them to be solved within my lifetime, ideally before I begin to chew the cud. This book is therefore partly an attempt to crowd source solutions to my own personal bugbears, all eighteen varieties of them.

Some science problems have relatively straightforward solutions, the demarcation line between solved and unsolved questions is clear cut, and standard scientific methods can be successfully used for their solution. In contrast, some research fields seem to constantly churn up the water around a problem, which provides little chance of a clear-cut solution and creates lots of muddy water. In some research areas it makes your head spin just trying to keep up with all the muddy water literature. Worse still, the muddy water is often a breeding ground for bull sharks that thrive in the opaque conditions and will readily attack anyone fearless enough to dip a toe into their precious unsolved problems. I do not want to denigrate people who stick at research on hard problems over the long term, particularly those who are consistently brave enough to try some new tools. However, it provides some satisfaction to give oneself the moral high ground when trying the patience of experts in a particular field and, if a problem has been known for fifty years without solution, there is every chance that another fifty years will slip by without any serious inroads being made. Hopefully this goal can be accomplished without too much mind numbing naivety and the lifespan of problems is always longer when they lack publicity. Solution to nearly all of the unsolved problems described in this book would provide large benefits to mankind.

An L next to the problems in the text stands for a lollipop and indicates that an important problem will be described. A lollipop will be donated to the first person to convince me they have solved each of these problems. Publication in a peer reviewed scientific journal is one of the prior requisites for consideration for these lollipop awards.

I should also make a quick disclaimer, since biological physics often overlaps with fields that are relevant to medical practice. Although I aim to correctly pinpoint the critical issues in the research described, blue skies research is always much riskier than would be acceptable in a clinical environment. Therefore, it goes without saying that none of the ideas in this book should be used to motivate unsupervised experiments on real patients. Clearly the help of health care professionals is needed for proper clinical experiments, both from a legal and an ethical perspective, following the guidelines of the corresponding regulatory bodies.

Each chapter is written as a self-standing piece, so they may be read in whatever order is preferred. The first sections of each chapter are intended for a general audience interested in some new science problems, not just for other physicists. These general introductions are then followed by a reading list and then a more advanced discussion is made for specialists who want to bridge the gap to specific research topics. There is thus a two-tier approach to the presentation; extended introductions for non-specialists followed by fairly compact technical details sections. An additional hope is that the technical details sections will allow readers to make contact with more of the underlying physics, losing some of the superficiality of a dedicated popular account, while not hampering the accessibility of the book. Nevertheless, non-specialists are advised to avoid the technical details sections on first reading due to the more advanced nature of some of the material. There is a companion book *The Physics of Living Processes*, also written by the current author, which is recommended for people who want a more detailed account of some modern physical tools for biophysical research, that might be useful for the solution of some of the problems described in the following pages.

Acknowledgements

I thank my family for their support (Roger, Sally, Cathy, Paul and my daughter Emily). Some of the work was completed at the Horizonte School in Regensburg who should be thanked for their hospitality. Furthermore, I should thank both my students and my colleagues in Manchester for making our School of Physics and Astronomy a pleasant environment in which to work, and their encouragement during the development of the manuscript.

All of the problems described in the eighteen chapters touch on my own research interests. I have profited immeasurably in these endeavours through discussions with colleagues. To acknowledge the help and advice given, my colleagues are listed in order of the chapters (science problems) to which they have provided input. However, I should make it clear that none of my colleagues should be held responsible for any technical mistakes I have made in what follows and many of them have much more expertise than I do with respect to some of the specific problems described.

Roger Waigh and Jeff Foreshaw should be thanked for providing their general opinions on the contents of the book and some editorial assistance.

Chapter 1. Viki Allan, Philip Woodman, Peter Weightman, James Hitchen and Jack McLean.

Chapter 2. Viki Allan, Philip Woodman, David Kenwright, Salman Rogers, Neftali Flores-Rodriguez, John Trinick, Tanniemola Liverpool, Cristina Marcetti, Emauela di Cola, Manlio Tassieri and Andrew Harrison.

Chapter 3. Rasmus Peterson

Chapter 4. Rasmus Peterson, Laetitia Adkins, Tom Flavell, Alistair Fensome and Trevell Paul.

Chapter 5. Mike Evans, Tanniemola Liverpool, Tom Mullin, Peter Stansby and Alex Malm.

Chapter 6. Dave Thornton, Gleb Yakubov, Alex Horsley, Paul Pudney, Pantelis Georgiadis, Aris Papagiannopoulos, Rama Bansil and Katrina Ribbeck.

Chapter 7. Jian Lu, Neville Boden, Amalia Aggeli, Alberto Sianni, Aline Miller, Hai Xu, Henry Cox, Peter Mawer and Lisa Carrick.

Chapter 8. Mark Dickinson, Alex Malm, Matthew Harvey, David Batchelder and Christoph Rau.

Chapter 9. Athene Donald, Mike Gidley, Dave Thornton, Sabine Fitsch, Alison Smith, Christian Riekel and Martine Debet.

Chapter 10. Ian Roberts, Chris van der Walle, Sally Waigh, Sorasak Phanphak, Johanna Blee, Jack Hart, Ruiheng Li, Douglas Kell and Jane King.

Chapter 12. Sergei Fedotov, Nicholai Korbei, Viki Allan, Salman Rogers, David Kenwright, Andrew Harrison and Amal Altjar.

Chapter 13. Manlio Tassieri, Mike Evans, Tanniemola Liverpool, Christoph Schmidt, Pantelis Georgiadis, Aris Papagianopoulos and Rama Bansil.

Chapter 14. Ji Chen Li, Roger Waigh, Neil Hunt and Mike Gidley.

Chapter 15. Phil Woodman, Viki Allan, Pantelis Georgiadis, James Sanders, Andrew Chung and Cynthia Li.

Chapter 16. Jian Lu, Eric Blair and Ian Hamley.

Chapter 17. Claudine Williams, Ralph Colby, Emanuela di Cola, Laura Fumagalli and Aris Papagiannopoulos.

Chapter 18. Henggui Zhang, Sanjay Kharchi and Michael Coleman.

Author biography

Tom Waigh

Dr Tom Waigh was born in Glasgow, but was brought up in South Manchester and went to Wilmslow County High School. He was a physics undergraduate at the University of Edinburgh (1990–94). He then completed a PhD in the Cavendish laboratory at the University of Cambridge. This was followed by a two year post-doc at the Collège de France in Paris in the laboratory of Pierre Giles de Gennes. He then returned to the UK with a lectureship in physics at the University of Leeds. Currently he is a senior lecturer in biological physics in the School of Physics and Astronomy at the University of Manchester. Previously he has written two books on biological physics, *Applied Biophysics* and *The Physics of Living Processes: a Mesoscopic Approach*, published by Wiley. He has published over eighty articles.

IOP Publishing

Some Critical Questions in Biological Physics
A guided tour around the bugbears
Thomas Andrew Waigh

Chapter 1

Molecular communication—crackling phone lines

1.1 Introduction

In terms of energy costs, molecular communication is a cheap method to transfer information in living creatures. Just emit a burst of a specific type of molecule somewhere in an organism (the *emitter*), let the molecules diffuse among the organism's cells and then detect this signal at another point in the body of the organism (the *receiver*). Alternatively, the molecule could diffuse through the air to be detected by another organism, where it acts as a pheromone. Moths are found to be extremely sensitive to pheromones and can detect single molecules. The subject of human pheromones is still hotly debated, but it is thought to be a minority mechanism of molecular communication, if it occurs at all.

Unfortunately, over large length scales such diffusive mechanisms for molecular communication, either internally or externally, are slow and fairly indiscriminate. Rather like someone yodelling all their personal details to a neighbouring alpine village, rather than sending an email to a relative's laptop. All the cells hear the same global signal regardless of their type and location, and it is hard to modulate the signal's intensity. This passive mechanism is used internally in humans for some slow globally important signals, such as those carried by hormones like testosterone, oestrogen and thyroxine. However, over short length and short time scales when molecules are confined to small specialized compartments, diffusive signalling can be very fast, efficient and well targeted, and it is consequently frequently used in such scenarios. For example, neurotransmitters travel across the extremely small distances (around thirty nanometres) in the synapses between neurons using diffusion. This is one of the factors that determine a person's speed of thought as they read these words.

An extremely general framework has been developed to describe the transmission of information between an emitter and a receiver by Claude Shannon. He was working at Bell laboratories in the USA during the 1940s when he developed his

doi:10.1088/978-0-7503-1374-2ch1 © IOP Publishing Ltd 2017

theory and a primary use was therefore in the telecommunications industry. In the telephone business it is critical to understand how noisy mechanisms of transmission degrade the quality of the transmitted signal during a phone conversation and Shannon's theory was very successful for the optimization of many of the contributing factors. There are therefore now advanced mathematical tools developed for telecommunications, and subsequently generalized in computer science for all processes of data transfer, that can also be used to calculate the rate at which information can be moved in biology and how it is affected by the levels of noise. The theory was originally used to understand how to control the crackling background noise on old fashioned analogue telephones, but these ideas have subsequently been applied to electrochemical signalling in the brain. This analysis of biological problems could be expanded even further. A hope is that a detailed understanding of the constraints on the transfer of information can provide important insights into a wide range of signalling processes in biology, e.g. biochemical signals that are implicated in cancer.

One important aspect of data transmission is bandwidth, i.e. the time response of the communication process (also an issue with internet speeds and *broadband* internet is the current solution). Often the bandwidth involves averaging over lots of noise at very short times scales and then there is an optimal time window over which the emitter and receiver can communicate, e.g. approximately 24 h for the internal clocks that drive the sleep/wake circadian cycles in humans. A better understanding of these time windows will be crucial for a fundamental understanding of the signalling processes in biology. This follows the vision of *cybernetics* created by Norbert Weiner in the 1940s who worked as a professor in mathematics at MIT in Boston. Cybernetics is the science of regulation (control) and communication in animals and machines. However, science has moved on since this era, with large advances in both computational research and molecular biology. The cybernetics approach has now been rebranded *systems biology* when applied to biochemical circuits. The reverse engineering of the biochemical circuits that occurs naturally in organisms is now a huge area of active research, within this field of systems biology, and the dynamics of these circuits plays a crucial role in the transmission of information.

L. *Question:* Calculate how the information rate is affected by noise in a range of biochemical circuits, measure their frequency responses (bandwidths), and quantitatively test the models that are developed.

For example, diffusive transport of neurotransmitters across synapses could be considered (already fairly advanced models exist here), thyroxine in the thyroid gland or adrenaline in the heart.

The regulation of intracellular transport also plays an important role in molecular communication. The time for a particle to diffuse the height of a giraffe in a nerve cell in its neck (a gigantic five metres in length) is on the order of thousands of years and similar impractically long times are presented to the nerve cells of other mammals, such as in ourselves, by the diffusive transport of particles. In terms of this specific example, neurotransmitters need to be constantly replenished at the end of nerves or the nerves can no longer communicate with their neighbours, which results in serious complications for the organism, e.g. this is a cause of motor neuron disease in humans (the mathematical physicist, Steven Hawking, is a famous sufferer from this disease and the comedian Ronnie Corbett was another). The evolutionary solution to this problem of intracellular motility has been to motorise the transport of the molecules. Protein nanomotors help to speed up the motion of neurotransmitters and replenish them at the ends of the nerve cells, but the motors also tend to increase the rate of information transfer when they are globally active, and facilitate transport in a wide range of other intracellular processes.

Metabolic regulation involves all of the processes in an organism's physiology, and it is also based on the transfer of information. How does a single cell in an organism that consists of billions of cells, know whether it should be alive or dead? Cells are not restricted to the rule, 'try to live for ever', and abnormal cells that occasionally follow this command often cause disease. For example, programmed cell death is an important part of morphogenesis, e.g. why most people have five fingers on their hands and not four or six. Overly long-lived cells can cause growth defects, and immortal cells can contribute to processes involved in cancers. Furthermore, how does the cell know whether to work harder or take it easy when involved in a collective task in tissue? A cell clearly needs to be told what to do and needs a reasonably constant supply of information. The processes of regulation are exceedingly subtle and intricate, but appear to be fundamental for a profound understanding of a wide range of living processes and their corresponding diseased states, such as the many forms of cancer.

The huge system of interlinked networks in the metabolism of an organism could either be set up like a giant intricate stack of playing cards or, conversely, it could contain much more robust well-regulated processes. The second possibility appears to be the obvious pragmatic choice and it is often considered to be an underlying design principle in biology; the standard processes observed in living organisms need to be robust, and thus stable, to a wide range of environmental perturbations. Otherwise the organism will fall foul of the theory of evolution and it can be easily outcompeted by other organisms that have more robust metabolisms. However, if a state of the art textbook on gene regulation is read, which contains the complete set of mechanisms used to read off information on people's DNA, there is seen to be a huge range of chemical possibilities and no obvious coherent collective regulatory features that decide between them; a metabolic stack of playing cards. Evidently there are significant gaps in our understanding of genetics (chapter 11). Furthermore, many of the chemical reactions involved in metabolic regulation need to be carefully compartmentalised and controlled, to stop adverse cross-reactions, and how robustness emerges from the transmission of all the associated information is not well understood.

It is hard to get to grips with the vast level of complexity involved in both intra- and intercellular communication in multi-cellular organisms, although it is expected that one day all the pieces of the jigsaw will click into place. Think of debugging a huge piece of complex software, which contains many hundreds of pages of code (e.g. all the background processes on a Windows machine) using a sophisticated programming tool. It is mindboggling to try to follow all the components of the code simultaneously, but if the code is steadily broken down using all the constraints on the different subsections, following the traditional reductionist method, the code's meaning could start to become evident. A much more detailed understanding of molecular communication in biology appears to be possible using this analogy and it will require extensive computational analysis.

For high speed communication over larger length scales inside organisms, where neither diffusion nor motor proteins are sufficiently fast, evolution has created mechanisms based on both analogue and digital electrochemistry. Analogue electrochemistry occurs for the fast transmission of signals in small animals (e.g. paramecium or insects) or at the start of a sensor's circuitry in larger animals. Paramecia are single-celled organisms that live in fresh water and are commonly sold in pet shops as food for tropical fish. They can be considered as an early link in the evolution of neuronal cells, since electrical potentials allow them to react quickly to environmental perturbations, e.g. when the single cell swims into an obstruction and it needs to reverse its motion. Digital electrochemistry is thought to be required in big brains and the bodies of larger organisms for improved efficiency (there is a lower energy cost) and it invariably occurs in such systems. Analogue signals involve continuously varying voltages in electrically excitable cells and are often found in sensory cells with humans (e.g. in the retina), whereas digital signals are carried by voltage spikes that are stereotyped (they all look the same) and they are used in other specialised types of electrically excitable cells that provide the longer ranged transport of information, e.g. all the neurons in the human brain. All living cells have a voltage across them due to imbalances in ion concentrations between the internal and external environments, but only electrically excitable cells have evolved to actively modulate these voltages and thus transfer information (see chapters 3 and 4). In humans the cells that use electricity to provide high speed communication are relatively small in number, but they form a vital fraction of the total number of cells in humans, and they use a disproportionate fraction of the body's total energy budget (over 20%).

Thousands of fans enjoyed Hugh Laurie's acting as Dr Gregory House in the American television series of the same name. His role in the fictional show was to troubleshoot the diagnosis of challenging diseases in patients; a last ditch attempt to cure desperate people who had exhausted all the other more conventional approaches. The scripts tended to involve the cast reeling off a huge list of exotic diseases every week, in search of an explanation for a particular patient's weird and wonderful symptoms. In reality, many of the diseases described in the show are indeed still poorly understood in modern science, particularly the subtle ones associated with signalling phenomena, such as cancers or autoimmune diseases like lupus.

Intracellular transport uses a packaging system to distribute signals to different compartments inside the cell and one class of the packages (fatty bags from 20 nm to microns in size) are called *endosomes*. *Exosomes* can also occur, where the fatty bags that contain signalling molecules are released into the blood system. They allow signalling over much longer distances than possible with endosomes and transmit information between cells. Returning to intracellular endosomes, their surfaces are found to be peppered with an array of signalling molecules, which act as postage stamps and addresses to determine the final destinations of the cargoes inside cells. For example, the signalling molecules might decide whether a package was due for complete degradation, whether it should be recycled with some minor reprocessing, or that it has an important message for the nucleus to be processed by gene networks. The surface molecules associated with signalling are often proteins that direct the postal service and they can be involved in analogue calculations with regard to the packages' fates. These calculations contribute to the behavioural decisions that are made by the cells.

One example of a labelling protein that occurs on endosomes is provided by the Rab family of transmembrane proteins. Rab is the diminutive form of Robert in Scots, so readers may start thinking of a Rab C Nesbitt soliloquy in Glasgow's George Square (a BBC comedy series starring Gregor Fisher). However, the reality is slightly less dramatic and the acronym stands for *Ra*t *B*rain, due to the system in which they were originally discovered (they are now known to occur in all human cells). Is it perhaps something Mary Nesbitt might call her husband after he spent a night on the tiles? A specific example, Rab5 (there are 66 varieties currently known, imaginatively named Rab1 to 66), was initially identified in a patient who suffered from lupus, and the current medical assumption is that it is somehow associated with the disease. Lupus is an autoimmune disease in which the immune system becomes overactive and attacks healthy tissue. A celebrity patient recently diagnosed with the disease is Selena Gomez (Justin Bieber's ex-girlfriend and a Disney channel actress). The disease can be fatal and its treatment is challenging.

When external chemical signals arrive on the walls of cells they can be internalised in small fatty droplets; the endosomes referred to previously. When they first arrive inside the cell the endosomes are called *early* and over time they coalesce, are molecularly modified, are transported by motor proteins, and finally they are called *late* endosomes. The *early* and *late* designations are intentionally vague, because the scientific story is not yet well established. Molecular markers for the *early* and *late* designations are Rab5 and Rab7, respectively. The conversion from Rab5 enriched endosomes to the Rab7 enriched equivalents is due to the interactions between the genes responsible for the two proteins in a chemical circuit and it has a switch-like function. The contents of early endosomes may be sent for reuse (high Rab5 concentration), whereas late endosomes are sent to the rubbish tip (high Rab7 concentration), where they are cut up into their component parts. The Rab5/Rab7 switch is involved in the decision as to which cargoes are designated *rubbish*. Clearly, if things go wrong and the switch malfunctions, perfectly good molecules can be needlessly sent to the rubbish dump or conversely rubbish (or even worse, pathogens, such as the influenza virus) can roam freely in the cells and start to disrupt their metabolism.

L. *Question*: Create an accurate quantitative model for the Rab5/Rab7 switch in the maturation of endosomes.

Currently there is only one sophisticated mathematical model in the literature for the Rab switch, which contains so many arbitrary assumptions that it is seen to be far from complete. The model either represents a broad brush stroked picture for a final accurate description or it may contain a significant number of errors. However, it is a good starting point and it acts as a strawman. The model is fairly easy to dispose of (any old spark will set the straw alight) and, although it has some clear weaknesses, it should attract more researchers to the field. For example, RNA regulation of genes (chapter 11) is at such an early stage of understanding that it is not clear what role it plays in the Rab5/Rab7 switch and the current model may be unrealistic in this respect. Genetic research in general still contains many fundamental unanswered questions.

In the calculations for the Rab switch, robustness provides a clear underlying principle to test any prospective models. In general, it is relatively straightforward to construct a number of exotic non-linear mathematical models for genetic switches, but if they are too sensitive to specific environmental factors the models will be invalidated. For example, slight changes in metabolites should provide broadly the same behaviour for the Rab switches inside cells, and the phenomena involved should not depend on the exact number of molecules present. A related concept is important in medicinal chemistry for drug delivery and is called the *therapeutic effect*. This is the drug dosage that causes toxic effects in 50% of the patients, divided by the dosage required to provide maximum therapeutic effects. Large therapeutic effects imply large ranges of dosages can be used with patients, which improves the safety of a drug. A 15th century quotation captures this as 'Everything is a poison, nothing is a poison. It is the dose that makes the poison'. Therefore, good drugs also do not require that exact numbers of molecules are used. Metabolic regulation that involves exact numbers of molecules is typically very hard to control in living systems and would rapidly result in an organism's death due to errors in counting. This theme of robustness can be used to whittle down the number of possible models for molecular communication, but there are still many unanswered questions in models for Rab switches.

Furthermore, there are other aspects to the Rab5/Rab7 phenomenon that require separate models. Aggregation models are needed to explain increases in the size and changes in the shape of endosomes as they mature from early to late stages of growth. Intermediate stage endosomes can adopt strange morphologies that resemble crop circles, as valuable cargoes pinch off the main spherical body of the endosomes in the form of tubules. The contents of the tubules are then recycled by the cell. For the infection of cells by viruses (such as influenza, which is contained within the endosomes), variation of the endosome's pH also plays an important role, which determines when the viruses can burst out of their endosomal packages and infect the cells. This crucial process for infectious disease is not understood with any quantitative detail.

There are hundreds of other equivalent unsolved problems in molecular signalling similar to the Rab5/Rab7 system, but this is a current known unknown. A fairly simple experimental barrier to the research is a lack of good antibodies to detect the presence of Rab7. There are commercial varieties of Rab7 antibody available, but our laboratory is yet to find one that is particularly successful for cell imaging. Antibodies can act as specific molecular glues to anchor fluorophores onto a particular protein and allow them to be used to track the proteins inside a cell using fluorescence microscopy. When better labels become available (Rab7 genetically labelled with a green fluorescent protein is also a good possibility, although these fluorescence labels are typically less bright than synthetic fluorescent probes, which compromises image quality) much more detailed tests of the mathematical models will become possible. Green fluorescent proteins are now one of the standard work horses of molecular biology. They were originally extracted from jellyfish and can in principle be used to label any protein molecule of interest in cellular biology using genetic expression (Nobel Prize 2008), as long as the position of the label does not disturb the function of the protein. Thus, fluorescence microscopy that allows simultaneous imaging of multi-coloured fluorophores is perhaps the most appealing experimental method to push the quantitative models forward, where the relative concentrations of Rab5 and Rab7 labelled with different coloured fluorophores could be followed in real time in live human cells, and it would help drive the development of better models.

In conclusion, many of the necessary tools, both experimental and theoretical, are available to solve lots of the problems in biological signalling, and great progress for biophysics research in this field is expected over the coming decade. These developments will in turn have a huge impact on our understanding of a diverse range of diseases caused by mistakes in information transfer, such as cancers, lupus and viral infections. Quantitative medicinal chemistry may also become possible, in which exact metabolic targets have been located based on the needs of information transfer, to help the development of new more effective drug treatments.

1.2 Suggested reading

Introductory

Gleick J 2012 *The Information* (London: Fourth Estate). Describes a wide ranging popular account of information theory, which includes historical aspects of Claude Shannon's life.

Sterling P and Laughlin S 2015 *Principles of Neural Design* (Cambridge, MA: MIT Press). Provides a fascinating semi-popular discussion of the designs of neural networks, with a good explanation of the strengths of analogue versus digital signalling in neurons.

Greenspan R J 2006 *An Introduction to Nervous Systems* (Cold Spring Harbor, NY: Cold Spring Harbor). A compact introduction to the neuronal networks that occur in simple organisms.

Advanced

Lim W, Meyer B and Pawson T 2015 *Cell Signalling: Principles and Mechanisms* (New York: Garland Science). A modern approach to the molecular mechanisms of signalling in biology, e.g. G protein cascades used to amplify light signals in the human retina.

del Vecchio D and Murray R M 2014 *Biomolecular Feedback Systems* (Princeton, NJ: Princeton University Press). Describes the application of engineering tools for feedback and bandwidth to problems in biology.

Nakano T and Ecford A W 2013 *Molecular Communication* (Cambridge: Cambridge University Press). This book provides a thought provoking discussion of the general framework necessary to understand the transmission of information using molecular communication. There is much work still to be done in this area to flesh out the bare bones of the problems presented, since biology has an extreme amount of detail pertinent to the issues involved in information transfer, and as yet many of the details are not well determined.

Atakan B 2014 *Molecular Communication and Nanonetworks: from Nature to Practical Systems* (Berlin: Springer). Similar to the Nakano book in scope, but from a more detailed mathematical perspective.

Del Conte-Zerial P, Brusch L, Rink J C, Collinet C, Kalaidzidis Y, Zerial M and Deutsch A 2008 Membrane identity and GTPase cascades regulated by toggle and cut-out switches *Molecular Systems Biology* **4** 206. Primary research literature that describes the mathematical models for Rab5/Rab7 signalling discussed in the text.

Foret L, Dawson J E, Villaenor R, Collinet C, Deutsch A, Brusch L, Zerial M, Kalaidzidis Y and Julicher F 2012 A general theoretical framework to infer endosomal network dynamics from quantitative image analysis *Current Biology* **22** 1381–90 Primary research literature that describes a mathematical model for the aggregation of maturing endosomes, e.g. aggregation of endosomes in the Rab5/Rab7 switch.

Alberts B, Johnson A, Lewis J, Raff M, Roberts K and Walter P 2014 *Molecular Biology of the Cell* (New York: Garland Science). Alberts is currently the standard text that describes the field of molecular biology. A detailed picture of the molecular biology of the cell is developing at a rapid pace, so the most modern edition is recommended.

Rieke F, Warland D, de Ruyter van Steveninck R and Bialek W 1997 *Spikes: Exploring the Neural Code* (Cambridge, MA: MIT Press). This is a very influential book that applies ideas from information theory to models of neurons and their activity in brains. However, the book is fairly mathematical and it is not recommended as an introduction for a general audience.

1.3 Technical details

Information theory is a sophisticated mathematical topic and here only a taster of a few results will be provided.

1.3.1 Definition of information

According to Claude E Shannon, information is a measure of surprise in a system and it is calculated as the logarithm of the probability of receiving a message, x ($p(x)$). An information entropy ($H(x)$) is defined mathematically as

$$H(x) = -\sum_{x \in X} p(x) \log p(x) \tag{1.1}$$

where x is an element of X, all the possible messages sent. The information entropy is thus maximised if all the messages are equally probable. This is a definition of information in terms of discrete variables, but it can be satisfactorily generalized to describe continuous signals (Stone 2015), such as those observed in spike trains that travel along neurons and the intracellular transport of signalling molecules.

1.3.2 Bandwidth calculation

Bandwidth is defined as the fastest rate at which a system can respond dynamically and in the context of information transfer it is given by the net bit rate, i.e. the number of bytes per second.

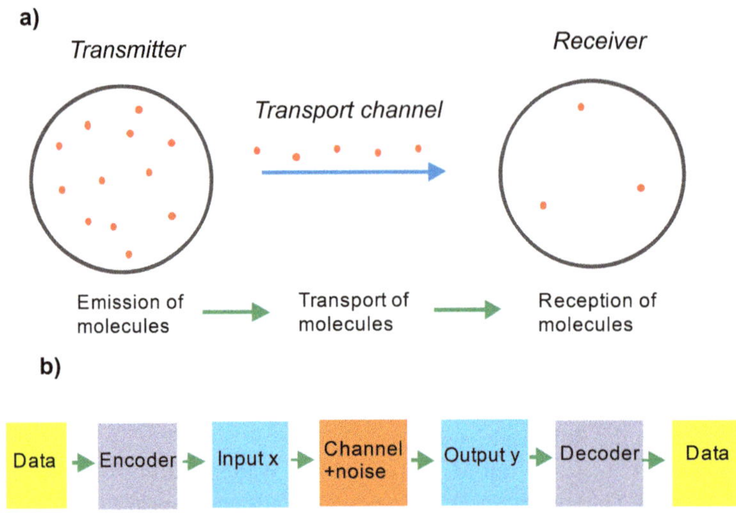

Figure 1.1. (a) General schematic diagram for the transmission of information in biological systems with molecules. Information travels from the transmitter to receiver through the emission of molecules. (b) Mathematically the molecular transport can be described by a process of encoding/decoding information.

1.3.3 Information theory

Consideration of the transport of information gives a fresh new perspective to understand the activity of biochemical signalling networks (figure 1.1). However, it is probably a secondary step that needs to be completed after traditional systems biology approaches. The challenge is to properly quantify the network architecture in the first place, before it can be combined with the dynamics of transport in these networks, to finally calculate the rate of information transfer. Thus, the rate of information transfer is a relatively high-level process that requires the networks to be well characterised beforehand, if the calculations are to be accurately performed. Nevertheless, the information theory approach does provide lots of interesting new hypotheses to explore the behaviour of molecular interactions, e.g. are the circuits measuring the concentration of signalling molecules or concentration gradients? Are the absolute amounts of the molecules or their release times being modulated to transmit information?

The mutual information ($I(x, y)$) of two variables x and y is the average information gained about y after a single value of x has been observed. Mathematically, it is defined for continuous variables as

$$I(x, y) = \int\int p(x, y)\log\frac{p(x, y)}{p(x)p(y)}\mathrm{d}x\mathrm{d}y \tag{1.2}$$

where $p(x)$ is the probability of x, $p(y)$ is the probability of y and $p(x, y)$ is the joint probability distribution of both x and y. More sophisticated theories can allow the calculation of the optimal rate of information transfer (related to how the signal is compressed and coded) (Stone 2015, Rieke *et al* 1997). There is a clear evolutionary advantage for high rates for information transfer and this constraint appears to be observed in many biological systems, e.g. it can lead to Weber's law in the human senses due to optimal signal compression (Waigh 2014).

1.3.4 The Rab switch model

A model for the Rab switch for the early to late transition in endosomes follows a standard systems biology approach (Del Conte-Zerial *et al* 2008, Ingalls 2013). Equations are written for the four key molecular species and their spatial variations are neglected, i.e. the reaction vessel is considered to be well mixed (which is certainly a crude approximation in endosomes that contain Rab proteins in lipid rafts on their surfaces). R_5 and r_5 are the concentrations of two molecular versions of Rab5, whereas R_7 and r_7 are the corresponding concentrations of Rab7.

Four coupled ordinary differential equations are needed to model the Rab5/Rab7 switch,

$$\frac{dr_5}{dt} = K_{-1} - (k_1 + GEF_5(R_5, R_7))r_5 + GAP_5(R_5, R_7)R_5 \tag{1.3}$$

$$\frac{dR_5}{dt} = GEF_5(R_5, R_7) - GAP_5(R_5, R_7)R_5 \tag{1.4}$$

$$\frac{dr_7}{dt} = K_{-2} - (k_2 + GEF_7(R_5, R_7))r_7 + GAP_7(R_5, R_7)R_7 \tag{1.5}$$

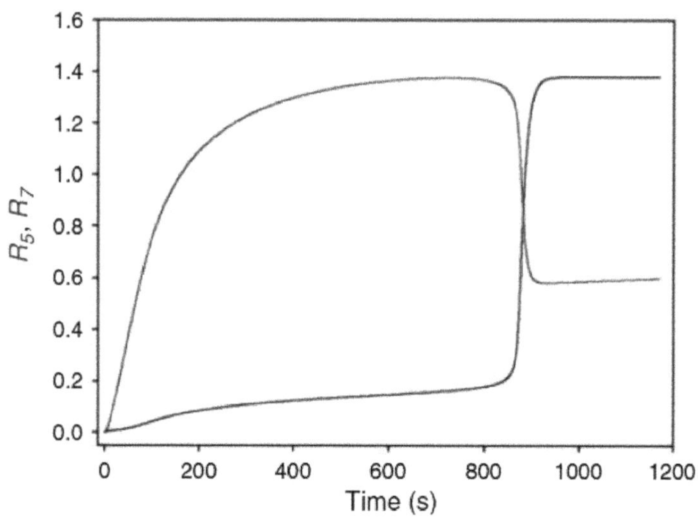

Figure 1.2. Rab5 (green) and Rab6 (blue) concentrations on the membranes of early endosomes as a function of time for the cut-out switch described by equations (1.3)–(1.6) (Del Conte-Zerial *et al* 2008).

$$\frac{dR_7}{dt} = GEF_7(R_5, R_7)r_7 - GAP_7(R_5, R_7)R_7 \tag{1.6}$$

where K_{-1} and K_{-2} are dissociation fluxes, k_1 and k_2 are association rates, GEF_i are activating functions and GAP_i are inhibitory functions. Some representative solutions to these equations are shown in figure 1.2, where the sudden non-linear switching behaviour of the R_5 and R_7 concentrations occurs at around 900 s and the cell would decide to send the endosome for degradation in its rubbish tip. A current challenge is to include more molecular details derived from experiments into this type of model, so that the model becomes predictive rather than just descriptive.

References

Del Conte-Zerial P, Brusch L, Rink J C, Collinet C, Kalaidzidis Y, Zerial M and Deutsch A 2008 Membrane identity and GTPase cascades regulated by toggle and cut-out switches *Mol. Syst. Biol.* **4** 206

Ingalls B P 2013 *Mathematical Modelling in Systems Biology: An Introduction* (Cambridge: MIT Press)

Rieke F, Warland D, de Ruyter van Steveninck R and Bialek W 1997 *Spikes: Exploring the Neural Code* (Cambridge, MA: MIT)

Stone J V 2015 *Information Theory: A Tutorial Introduction* (Sheffield: Sebtel Press)

Waigh T A 2014 *The Physics of Living Processes: A Mesoscopic Approach* (Chichester: Wiley)

Chapter 2

How dynein works and other motor proteins

2.1 Introduction

Inside human cells there are three main types of nanomotor, which move around cargoes that require high speed transport over reasonably long length scales, e.g. across the whole cell. Diffusion, in contrast, is a default type of motion that exists in all materials due to thermal jiggling that leads to slow transport and it is most noticeable with small particles when they are observed at small length scales. The nanomotors move at much higher rates than possible with randomized diffusive motions. The active driven motions by nanomotors are found to be critically important for the cells' wellbeing, since they carry many vital packages that need to be replenished at reasonably fast rates. Other crucial roles for motor proteins occur in specialized cell types, e.g. contractions of the heart are driven by huge parallelized arrays of nanomotors in dedicated cardiac cells (myocytes).

To put it bluntly, without the nanomotors, intracellular transport is too slow, none of the important packages arrive in time, and the organisms will quickly die. The tiny motorised devices that have evolved in human cells are all based on proteins, and the main varieties for intracellular transport are called *dyneins*, *kinesins* and *myosins*. The first two types of motor walk on microtubules (molecular railroads that are predominantly directed to and from the nucleus of a cell), whereas the third walks on actin filaments (a different variety of railroad, often found around the edge of a cell).

There is a much wider variety of intracellular machines in humans that also use ATP (the standard intracellular fuel), which have not been discussed, such as enzymes that drive DNA replication, those that create ATP in the first place in mitochondria (ATPases in the cell's powerhouse) and those that can manipulate the topology of individual strands of nucleic acid. However, these motors do not directly lead to significant amounts of intracellular transport and are ignored here. Furthermore, in evolutionary terms, bacteria have been extremely inventive in the

doi:10.1088/978-0-7503-1374-2ch2 © IOP Publishing Ltd 2017

creation of new mechanisms of motility and new ones continue to be discovered, but they will not be covered in detail in this chapter.

The three types of nanomotor highlighted from human cells, dyneins, kinesins and myosins, are all stepper motors. Each example of these molecular motors can use up one molecule of fuel (ATP) to cause one of their two legs to swing forward, to advance themselves along the molecular railroads on which they travel e.g. steps of eight nanometers are taken by dyneins along microtubules. Stepping at the nanoscale has some counter-intuitive aspects, since the thermal forces in the local environments of the nanomotors are relatively large and cause huge fluctuations in their positions. The motion of a vesicle (a small fatty package) attached to a dynein nanomotor in a cell is analogous to someone (the motor) walking with a helium balloon (the vesicle) in a hurricane; the balloon races around wildly, straining on its tether. Even stranger is the absence of inertial forces at the nanoscale, which is equivalent to a macroscopic walker trying to propel their way through a swimming pool filled with treacle. This is called the problem of low Reynolds number dynamics and it is also relevant to calculations of the transitions of fluid flows to turbulent states (chapter 5). Standard swimming strokes (reciprocal mechanisms, such as the breast stroke, the butterfly or the front crawl) would just cause the swimmer to oscillate backward and forward with no cumulative forward progress. Nature has thus set some large challenges to motors for them to achieve efficient transport at the nanoscale.

Kinesins and myosins are relatively well understood nanomotors. This statement more reflects our lack of understanding of dynein, rather than any expectation that there are no significant gaps in our knowledge of the other two types of motor; there certainly are. The main challenges with dynein are that it is a bigger molecule and thus more complicated than kinesin or myosin, and it also has more accessory proteins. These accessory proteins, given strange names, such as BicD2, NudE, NudEL, LIS1 and dynactin, are vitally important for dynein's activity, but they are hard to assemble in the correct order outside the cell, which is the standard experimental method used to simplify matters to develop instructive structure–function relationships. Dynein nanomotors are thus badly understood. Here the subject of cytoplasmic dynein is being discussed, i.e. intracellular nanomotors in human cells. A different class of dyneins, called axonemal dyneins, are found in cilia, but they have different questions associated with their activity, e.g. how they perform their jobs as a dust escalator in the lungs.

A number of roles are postulated for the accessory proteins that accompany cytoplasmic dyneins, which includes latches, springs, gears, signals and clutches. No one yet really knows what they all do together. A human form of motor neuron disease occurs if one of dynein's accessory proteins (dynactin) malfunctions, so clearly they have important duties. Dynactin is thought to be necessary for dynein's processivity. Without dynactin, the dynein molecules are statistically more likely to drop off the microtubular railroads. Thus, without dynactins the motors travel less far, compromising the activity of the transport networks, which fail to recycle molecules from neural junctions, compromising their activity. The neural junctions in turn are needed to transmit information between neurons, and paralysis and sensory deprivation result when they are lost.

L. *Question*: Provide a molecular model for dynein motility inside a live cell, which includes the activity of all of the accessory proteins, such as BicD2, dynactin, NudE, NudEL and LIS1.

In general, attempts to reconstruct the activity of intracellular motors *in vitro* outside the cell find that they often run slow, if they move at all. Assuming all the biochemistry has been correctly performed, this is surprising, since there is so much stuff inside a cell that it might be expected that a motor protein would need much more force to push its way through the obstructing material, causing it to slow down i.e. intracellular viscosity is much higher than the water used *in vitro* and the increased viscosity inside a cell will resist any motion. This is often the opposite trend to that seen experimentally.

L. *Question*: Quantitatively explain why motor proteins sometimes go faster *in vivo* for the transport of cargoes in human cells than *in vitro*.

The molecular processes that determine protein motility are not statistically independent and have a memory due to caging behaviour and their statistics are called non-Markovian. One possible explanation for this behaviour of motor proteins is therefore the non-Markovian nature of the binding of fuel molecules to nanomotors, e.g. the addition of fuel molecules to the motors occurs differently in crowded environments and their consecutive displacements are strongly correlated. The motors inside cells act in a dense suspension of other molecules and they can thus have a memory of their previous movements. If a particle bounces off the wall of an obstruction in a densely packed system, the statistics of its motion have a memory of its collisions. Does this crowding effect increase or decrease the activity of the motors? It may well increase the activity, since the crowded environment could increase the uptake of fuel, which results in the experimentally observed fact that motor protein rates are often faster *in vivo* than *in vitro*. However, this is currently just a speculative theory and needs to be rigorously tested with carefully performed experiments. Non-Markovian effects certainly do occur inside cells, but their impact on reaction kinetics, and thus the speed at which motors can run, is not well understood.

A simple technical issue with motor proteins involves problems with tracking particles using fluorescence microscopes. As fluorescence cameras become able to acquire data faster, so too do the reported motor speeds in the literature, and these speeds are consequently slowly creeping up with time. This is not necessarily due to trivial misanalysis of the data, although mistracking of fast particles is always an issue. At short times the motion of nanomotors is a combination of diffusive motion and motion directed by nanomotors. Diffusive motion is fractal-like (chapter 12) and has the peculiar property that the velocity calculated depends on the time scale at which it is measured. Naïve analysis of motor protein velocities can therefore just capture this unusual diffusive behaviour, with little information provided about the underlying action of the motors. There are therefore many question marks with regard to the quantitative values currently provided in the motor protein literature for parameters that describe their motility, such as their top speed. As camera technologies improve in combination with the resolutions of microscopes, the photochemistry of the fluorescent labels and the mathematical models used to describe the experiments, it is hoped that a more robust picture will be developed for motor protein motility. An additional complication is that diffusion inside cells has some unusual statistics (it is called sub-diffusive), which needs the development of fairly sophisticated mathematical tools for its description (chapter 12).

The field of thermodynamics was developed at the same time as the industrial revolution in the 1800s. Inventers needed models to guide the construction of new designs for steam powered engines that were fuelled with coal. The thermodynamic models that were created provided a useful deterrent against crackpots who claimed to have made engines that did not require any fuel and they helped to optimise the performance of other functional prototypes. The thermodynamic framework that was created is of such a general nature that it is equally applicable to biological motors at the nanoscale as to steam engines in full sized locomotives. Often biologists are unfamiliar with the application of thermodynamic principles and occasionally claims can be read in the literature of the rectification of Brownian motion in protein nanomotors, which leads to unfuelled motility. At first sight, these claims may not appear to be too crazy, but more careful analysis implies the complete failure of the second law of thermodynamics in these cases and the resultant creation of a perpetual motion machine, i.e. a motor that runs without any fuel. Such occurrences are the stuff of nightmares for card carrying physicists (they are contrary to a fundamental law of nature), and these studies need to be carefully revisited to discover the source of the errors.

Single nanomotors are interesting entities, but with large collections of them the range of possible behaviours becomes even richer. Consider the intricate molecular choreography of the millions of muscle proteins in the fingers of a pianist who plays a Bill Evans solo or starlings that flock on a city skyline (a murmuration). Another example is the swarming behaviour of thousands of self-propelled synthetic nanoparticles, and it is currently a very active area of research, both theoretically and experimentally. The research on the collective motion of motors is also relevant to many medically relevant systems, such as clouds of mosquitos that carry malaria or infected wounds that teem with deadly bacteria.

These shoaling phenomena of flocks of motile agents show a curious interplay between the underlying internal motion of the nanomotors and the hydrodynamics of the surrounding environments, and some fundamental insights have been driven by the need for accurate computer animations in the movie industry. Thus, amusingly, the algorithms constructed for movies, such as *Finding Nemo*, have given rise to new physical models that could in turn be used to study the animals (e.g. tropical fish, mosquitos and starlings) depicted in these films in the wild. Specifically, Vicsek type models of the swarming behaviour of active matter are thought to have their origins in the animation industry.

Liquid helium demonstrates the fascinating phenomenon of superfluidity. Drop the temperature to below an impressively cold 2.172 Kelvin (−270.979 °C) and give the sample container a shake, and the flows induced in the fluid will in principle continue indefinitely, since there is no dissipation. In fact, the creation of vortices (plug-hole type motions) in a superfluid can dissipate a little energy and the flow does not continue forever, but the liquid does have zero viscosity at long time scales. This behaviour is well described by the quantum mechanics of the helium atoms at low temperatures, since they act as Bose particles (Bosons) and the exotic phenomena involved with these helium superfluids have been the subject of four Nobel prize winners' research. The Bosons experience negligible inter-particle interactions, which leads to negligible frictional coefficients and thus the materials have zero viscosities. More surprising is that suspensions of bacteria can have zero viscosities too. Negative viscosities are even possible with bacteria, if they choose to swim strongly in the opposite direction to that of the shearing force. However, the origin of the superfluidity in bacteria is completely different to that with liquid helium, since no quantum Bose particles are directly involved. The origins of bacterial viscosity are found in the activity of propellers on the microorganisms and their tendency for mutual alignment.

L. *Question*: Make an accurate quantitative model for superfluidity in suspensions of swimming microorganisms.

An averaged (mean field type) model for the bacterial velocities is required to describe the superfluid suspensions that have zero viscosity. Strange non-linear phenomena associated with velocity fluctuations are also possible in bacterial suspensions, dubbed *bacterial turbulence* (distinct from conventional classical turbulence that is also badly understood), which are much more challenging to describe theoretically (chapter 5). Richard Feynmann called *turbulence* the most important unsolved problem in classical physics. *Bacterial turbulence* and *elastic turbulence* are two of classical turbulence's even weirder cousins. Bacterial turbulence is a robust repeatable phenomenon and, in common with classical turbulence, the velocity statistics of bacterial turbulence are intermittent for the bacteria, i.e. the velocities of bacterial suspensions erratically decrease and then increase again with characteristic fat-tailed probability distributions. Strong non-linearities in the systems cause this intermittency in both bacterial and classical turbulence. Nevertheless, the hope is that bacterial turbulence can be ignored in the search for good quantitative theories of the superfluidity of bacteria, since it poses such a formidable barrier.

Another enticing idea is that the design principles of nanomotors can be borrowed from biology and incorporated into synthetic systems. Nanotechnology applications would be possible where millions of autonomous motorised entities clean up oil slicks, work on production lines or scrub the insides of diseased blood vessels in stroke victims. Organisation of the collective behaviour of nanomotors is particularly well controlled in biology and it would be a significant advance if synthetic nanotechnology could learn this trick of nanomotor choreography. However, these are long term goals for nanomotor research, and in the medium term, an improved understanding of protein nanomotors should play an important role in the prevention of many of their associated medical conditions, such as motor neuron disease or cardiac disease in its multiple forms (chapters 1 and 18).

2.2 Suggested reading

Introductory

Hoffmann P M 2012 *Life's Ratchet: How Molecular Machines Extract Order from Chaos* (New York: Basic Books). A fascinating popular account of the physics of motor proteins.

Vogel S 2003 *Prime Mover: A Natural History of Muscle* (London: Norton). The book describes aspects of muscle research from a predominantly medical perspective.

Advanced

Bray D 2000 *Cell Movements* (New York: Garland). Lots of examples of nanomotors in zoology are provided.

Guigas G and Weiss M 2008 Sampling the cell with anomalous diffusion—the discovery of slowness *Bioph.* J. **94** 90–4. Describes the impact of subdiffusive kinetics on reaction rates inside cells.

Lopez H M, Gachelin J, Douarche C, Auradou H and Clement E 2015 Turning bacterial suspensions into superfluids *Phys. Rev. Lett.* **115** 28301. A primary research article that describes how bacterial suspensions can act as superfluids with zero viscosity.

Dunkel J and Heidenreich S. *et al* 2013 Fluid dynamics of bacterial turbulence *Phys. Rev. Lett.* **110** 228102. A primary research article that presents joint experimental and theoretical results on bacterial turbulence in *B. subtilis* bacteria.

Malm A and Waigh T A 2017 Elastic turbulence in entangled semi-dilute DNA solutions measured with optical coherence tomography velocimetry *Sci. Rep.* **7** 1186. A primary research article on turbulence in reasonably concentrated solutions of DNA.

Howard J 2001 *Mechanics of Motor Proteins and the Cytoskeleton* (Sunderland, MA: Sinauer). A classic introduction to the physics of motor proteins.

2.3 Technical details

2.3.1 Nanomotors

Dynein, myosin and kinesin (figure 2.1(a)) all act as nanomotors for the transport of cargoes inside cells and they are all stepper motors. ATP molecules (fuel) are used by the motors and a reaction cycle based on ATP causes one of the motors legs to move forward (figure 2.1(b)). A continuous supply of ATP can therefore cause continuous stepping motion of the motors in a constant direction. Understanding the exact

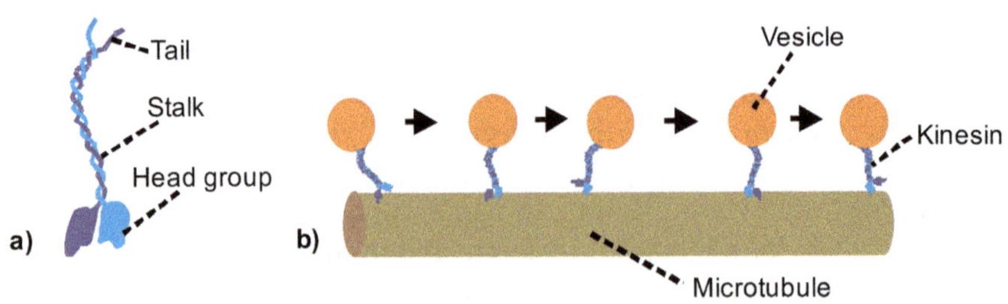

Figure 2.1. (a) A single kinesin molecule consists of a tail region that attaches to vesicles (the cargoes), a stalk spacer region and a head group that coverts ATP into step-like motion. (b) The walking motion of a kinesin molecule along a microtubule track.

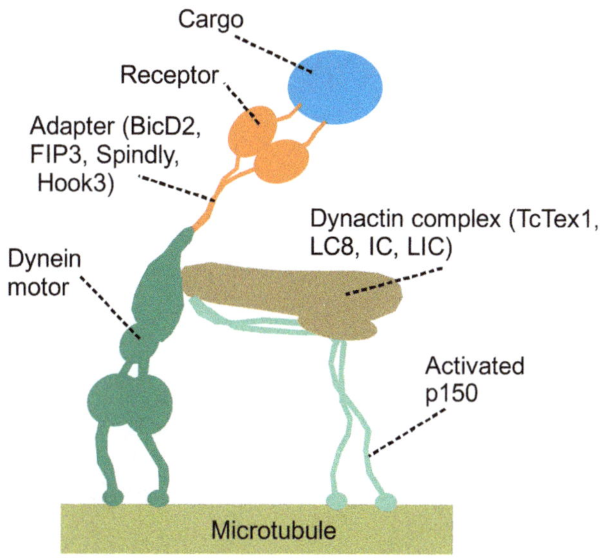

Figure 2.2. Dynein also has a head group domain similar to kinesin (figure 2.1(a)) that allows it to walk along microtubules. It also has a series of accessory proteins whose activity is not well understood, e.g. where NudE, NudEL and LIS1 are located (Kardon and Vale 2009, Allan 2014).

molecular mechanisms involved is tricky and there still remain many unresolved questions.

Cytoplasmic dynein is still a poorly understood nanomotor due to its large size and the many accessory proteins needed for its optimal performance (figure 2.2), e.g. BicD2, dynactin, NudEL and NudE. Some crystalline structures of the purified accessory proteins are available, but the exact functional roles of all the proteins *in vivo* are not well understood. The dynein motor unit is needed for motility to occur, but it only appears to use up ATP when the Zimmer frame structure (a dynactin complex with p150) is also present.

2.3.2 Vicsek model for shoaling of nanomotors

To understand the swarming behaviour of nanomotors Vicsek models can be used. These models have been used to simulate the motion of shoals of fish and the collective motion of many other animals (Vicsek *et al* 1995). The models are thought to have their origins in computer animations.

Computationally the main rule is fairly simple: *at each time step, a given particle is driven with a constant absolute velocity and it assumes an average direction of motion of the particles in its neighbourhood, of radius r, with some random perturbation added.*

Mathematically the position of each fish (x_i) at time, $t + 1$, is incrementally dependent on its position and velocity at the previous time step (t),

$$x_i(t + 1) = x_i(t) + v_i(t)\Delta t \tag{2.1}$$

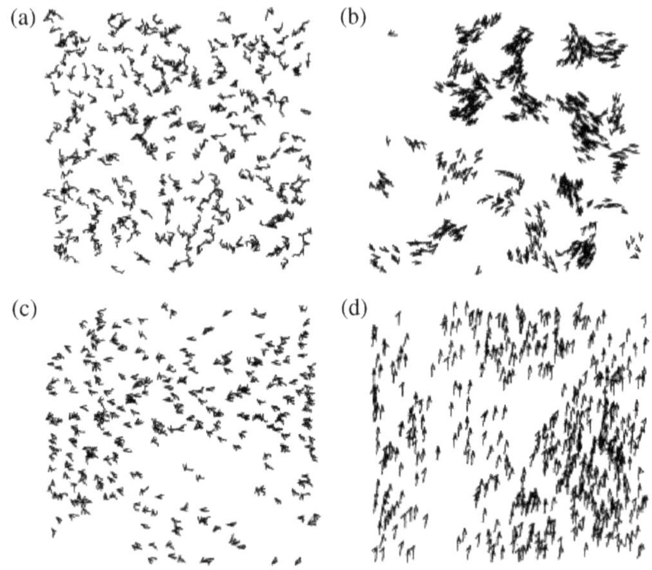

Figure 2.3. Velocities of the swimmers are shown for varying densities and degrees of noise in the system. For details see the original publication (Vicsek *et al* 1995). The emergent phases include (a) a random distribution (low density and no noise), (b) formation of small coherent flocks (low density and small noise), (c) random motion with some correlations (high density and high noise) and (d) macroscopic ordering and a single spontaneously selected direction (high density and small noise). Reprinted figure with permission from Vicsek *et al* 1995. Copyright 1995 by the American Physical Society.

where $v_i(t)$ is the velocity of fish i at time t. Furthermore, the direction in which the fish swim is incrementally updated based on the average direction of the shoal (it is assumed that the fish can monitor the direction of their neighbours)

$$\theta(t+1) = \langle\theta(t)\rangle_r + \Delta\theta \tag{2.2}$$

where $\langle\theta(t)\rangle_r$ is the average direction of the velocity of particles within a circle (radius r) surrounding a given particle, i.e. the localized shoal swimming direction, and $\Delta\theta$ is a random angular offset chosen from a Gaussian distribution.

Clearly, this is a simple model that is missing many physically relevant details, e.g. excluded volume interactions are needed to stop the fish swimming through each other like ghosts and hydrodynamic interactions between the neighbouring fish should also be included, since they perturb their motion. However, the important physical insight is that the Viscek model can lead to a dynamic phase transition in which spontaneous coherence occurs that is accompanied by flocking behaviour (figure 2.3). Similar coherent motile behaviour occurs in many natural systems, e.g. the motility of bacteria, starlings and geese.

2.3.3 Viscosity in bacterial suspensions—experimental data

The rheology measurements that have been performed on active swimming bacteria, in practical terms, are relatively straightforward and a representative data set is shown in figure 2.4 (Lopez *et al* 2015). The main challenge in such experiments is to

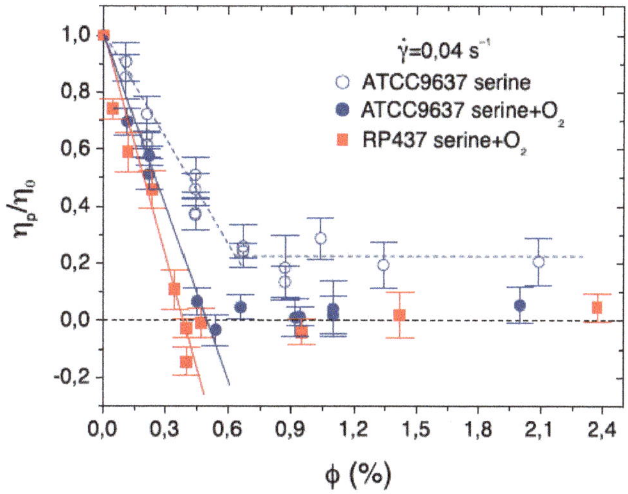

Figure 2.4. The viscosity (η_a) of bacteria (strains of *E. coli*) from non-linear rheology experiments as a function of bacterial volume fraction (ϕ). Superfluidity (negligible viscosity) is observed at high volume fractions of bacteria (ϕ>0.4) and it depends on the strain of *E. coli* used. Reprinted figure with permission from Lopez *et al* 2015. Copyright 2015 by the American Physical Society.

control the metabolism of the bacteria, so their motility is sufficiently well defined, e.g. the temperature, the surrounding nutrient broth and the amount of oxygen can all play a role in how they swim. Evidently, at higher volume fractions the viscosity can drop to zero or it can even adopt negative values. To model this behaviour, on the level of single bacteria, would require large numbers of them to be tracked. Some assumptions would then need to be made on their tendency for mutual alignment and their mechanism of flagellar driven motility. This is a very sophisticated problem in collective colloidal hydrodynamics, but it could be industrially relevant, e.g. to describe the optimal mixing of bacteria in bioreactors or the activity of nanomotors in lubricated bearings with zero viscosity (whether the energy budget of the bacteria would be higher than that due to the frictional loses in the bearing is open for debate).

References

Allan V J 2014 One, two, three, cytoplasmic dyenin is go! *Science* **345** 271

Kardon J R and Vale R D 2009 Regulators of the cytoplasmic dynein motor *Nature Rev. Cell. Biol.* **10** 854–65

Lopez H M, Gachelin J, Douarche C, Auradou H and Clement E 2015 Turning bacterial suspensions into superfluids *Phys. Rev. Lett.* **115** 28301

Vicsek T, Czirok A, Ben-Jacob E, Cohen I and Shochet O 1995 Novel type of phase transitions in a system of self-driven particles *Phys. Rev. Lett.* **75** 1226–9

IOP Publishing

Some Critical Questions in Biological Physics
A guided tour around the bugbears
Thomas Andrew Waigh

Chapter 3

How brains work—wiring and consciousness

3.1 Introduction

The functional mechanisms of the brain constitute the crown jewels of all the unsolved problems in science. Arguably a solution to this set of problems will have greater significance for the human race than those to any other question currently under investigation in science, e.g. it will be more important than knowledge of the origin of dark matter or a unified theory of gravity with quantum fields. The most sophisticated object as yet identified in the known Universe is the human brain, not a black hole or a Higgs Boson, which are thought to be relatively simple in comparison, slavishly following their basic underlying physical laws. Arguably the World Wide Web is an even more complicated entity than a single brain, but it only reflects the consciousness and intelligence of the brains used to create it, and as yet has not displayed these emergent phenomena itself. Furthermore, this is not to say that research into black holes and Bosons are without their merits, that they are completely well understood or that more sophisticated emergent phenomena will not be discovered at some point in the future. However, many of the finest minds in history have pondered issues involved with the brain, with some making public their findings, but still both consciousness and intelligence lack a rigorous detailed understanding, unlike Bosons and black holes. It is suspected that the vast majority of scientists and human beings in general, will have pondered to some extent what makes their own brains tick. How can progress be made with such a hard problem?

A pragmatic approach to understand how brains work is that of reverse engineering; in which a functional brain is considered and an understanding is developed in terms of its component parts. A perspective is that in the long term this will be the most successful approach, and the problem will eventually yield to the pressure of thousands of researchers' efforts to understand the miracles of consciousness and intelligence. It is now clear that these phenomena are due to the interactions of one hundred billion neuronal cells in the human brain. Furthermore, such miraculous emergent phenomena occur every day in the seven billion people that

doi:10.1088/978-0-7503-1374-2ch3 © IOP Publishing Ltd 2017

currently populate planet Earth, so they are evidently robust, repeatable phenomena and the author is convinced that they occur to some extent in his very own brain.

From the perspective of biological physics, it is felt that putting numbers to problems in brain research will help move it forward and thus should be encouraged. Quantitative models are therefore needed for the activity of brains, to really feel completely secure that how they work is properly understood. Just naming the anatomy, the cellular architectures and molecules in brains, on its own is not a very satisfactory approach in the long run, although it is an important first step.

The equivalent in electrophysiology of the Schrödinger equation used in quantum mechanics (chapter 14) is the Hodgkin–Huxley equations, and the discovery of these equations resulted in a Nobel Prize for medicine in 1963. Hodgkin and Huxley used neurons from giant squids to test their models due to their large size (around one metre in length and one millimetre in thickness), which facilitated electrophysiological measurements, e.g. early voltage clamp experiments were possible. The Hodgkin–Huxley equations describe the electrical activity of neurons in terms of the conductivity of the molecular ion channels embedded in the external membranes of the cells. The model can provide predictions for many of the experimentally relevant quantities, such as the speed of electrical waves that are conducted along neurons, the form of the electrical potentials (spikes) and how the voltages depend on the exact ion channels involved. For example, it is possible to correlate mutations in the genes for ion channels with the occurrence of diseases in hearts and brains. Thus predictive theories for the electrical activity of single neurons are fairly well developed, although inclusion of heterogeneous conductivity inside cells at the nanoscale still provides some challenges (chapter 18).

For progress to be made, the field of brain research needs to be framed in terms of tractable sub problems or researchers will be completely overwhelmed by the scale of the problem. The complete wiring diagram of the neurons inside the human brain when it is complete will be huge, since on the order of one hundred trillion connections need to be mapped for all the neurons relative to one another, i.e. one hundred billion neurons exist in a brain and each has on average a thousand connections with its neighbours. Nevertheless, with the advent of modern high-performance computers and tomographic electron microscopy, brain mapping is now in principle a tractable problem. Just slice and dice a human brain up into micron thickness slivers using a diamond knife, take an electron microscopy image of each of them at nanometer resolution and then use a computer to fuse all the images together into a single three-dimensional reconstruction of an entire brain. *Voilà*, a three-dimensional map of the brain's neuronal wiring diagram has been created. This is simple to say, but technologically it is very tricky to do. Indeed, such a project has recently received investment of over 38 million dollars in the USA under the guise of the *human connectome project*. There is also a European project currently underway to create a virtual brain that attracted one billion Euros of funding; a huge amount of money. However, the worry from a biophysics perspective is that the European project is predominantly a computer science vehicle that will result in lots of interesting CompSci, but may come no closer to providing answers to the critical biological questions on what makes people's brains tick. Pure theory without strong investment in experiment always has some clear weaknesses, so there is more optimism that new biophysics will be found in the human connectome project and its future extensions. Without real world experiments, research can stop being science in the Popperian sense, i.e. *A theory in the empirical sciences can never be proven, but it can be falsified, meaning that it can and should be tested with experiments.* Therefore, the human connectome project is included as a first tractable problem towards a quantitative understanding of how brains work.

L. *Question:* Create an accurate wiring diagram of a human brain that shows the position of every synapse.

Once this map of the brain is complete (and we are neglecting the fact that the strengths of synaptic connections play an important role and they cannot be deduced from the network architecture), the creation of *in silico* models for brain activity are appealing. Nevertheless, these models need to be bench-marked against real electrophysiological experiments or they risk becoming an abstract CompSci project, rather than proper biological physics. In the meantime, scientists are building up to the completion of such huge targets as the full human connectome project by aiming at slightly more modest goals, and provisional maps of the mouse brain have already been published.

Next the problem of consciousness will be considered, but the question will be couched in terms of some well-defined jargon from the theory of thermodynamics.

L. *Question:* Quantitatively describe the nature of the phase transition between conscious and unconscious states of the human brain.

Historically the field of thermodynamics was developed in the 1800s to describe how heat is transferred between systems, but it has subsequently been generalised to describe any change in the state of a material, i.e. phase changes of materials, such as the solid-to-liquid transition (melting of an ice cube) or the liquid-to-gas transition (boiling of a kettle) can be described. Classically, to describe a phase transition in thermodynamics a good parameter needs to be chosen. Ideally the parameter has a value of zero in the disordered state and a value of one in the more ordered state. For example, if water is boiled in a kettle, its density is a good (order) parameter. In the liquid state the order parameter can be set equal to one (e.g. the density rescaled by the original density at room temperature) and the density drops dramatically in the gaseous phase when steam is produced. With an ice cube an alternative order parameter might be a component of the crystalline pattern formed in the frozen state, e.g. the distance between two oxygen atoms. A wide range of other order parameters are found to be useful with other phase transitions, e.g. the electrical conductivity of metals, molecular orientation in liquid crystals or the local concentration of the components during liquid–liquid phase separation to describe vinaigrette recipes when oil mixes with water. Once the order parameter has been identified, a decision can then be made as to whether a discontinuous or a continuous phase transition has occurred (i.e. whether the phase transition occurs in an abrupt fashion) and some very general theoretical statements can then be made that describe the processes involved.

So, what is a good order parameter for consciousness? The blood oxygen level measured in fMRI (functional Magnetic Resonance Imaging techniques, such as

BOLD) experiments might be a good choice. BOLD imaging stands for blood-oxygen-level dependent imaging. A type of electrical wave in the cranial activity measured by electroencephalography (EEG) is another possibility. For example, you could place an array of electrodes around someone's head and measure the voltages created by their brain when they fall asleep. Large changes are seen in the EEG voltages dependent on the brain's activity, e.g. an alpha wave occurs if someone shuts their eyes. Other brain imaging devices, such as Doppler optical coherence tomography (OCT), positron emission tomography (PET) and ultrasound imaging, might also have good order parameters for consciousness associated with their measurements. A challenge is for the parameters to be sufficiently specific to conscious mental activity and that they are not subsumed by the background noise due to all the other jobs that the brain performs. The measurements therefore require high biochemical selectivity, high spatial resolution and high temporal resolution, without them destroying parts of the brain that are studied (chapter 8).

Consciousness and intelligence are thought to be two separate phenomena in the human brain, e.g. there is good evidence for the phenomenon of subconscious intelligence (the unconscious solution of problems faced by an individual) and the search light of consciousness appears to be an experimentally verifiable process in MRI experiments. Furthermore, some initial MRI work by a reputable group from Paris (at the Collège de France) already claims consciousness is a continuous phase transition. A popular account of the MRI experiments can be found in the book *Consciousness and the Brain* by S Dehaene. The researchers believe consciousness has more in common with the alignment of spins in a magnet (a continuous phase transition), than the abrupt changes that are more frequently found in nature, such as the melting of an ice cube (an abrupt first-order phase transition). Additional research is required to achieve a consensus of opinion over a broad community with such subtle experiments and reference should be made to details in the original literature. Nevertheless, there is a good chance an unequivocal answer will eventually be reached in the field and the arguments of the group from the Collège de France are currently very persuasive.

So what, if consciousness can be understood slightly better? Well, consciousness is involved with a number of important physiological switches, such as the sleep process. Many artificial switches have also been discovered by mankind through history for consciousness, which include alcohol and other narcotics, but also medically relevant drug molecules, such as anaesthetics are important. Research on how consciousness is switched on and off could be very useful in medicine.

L. *Question:* Describe quantitatively what happens to the phase transition involved in consciousness in the human brain upon the application of a general anaesthetic.

These studies of consciousness could also be extended to the exploration of brain disease. For example, sleep patterns are altered in many mental diseases, such as those observed with Alzheimer's patients, and the brain damage involved can perturb processes involved in consciousness. Thus, sleep patterns can be used as a cheap method to diagnose the progress of Alzheimer's disease.

There are whole undergraduate courses dedicated to neuroscience, which underlines the huge amount of information already known on the subject, and physicists do well to respect the large edifice of knowledge that is already in place. However, hopefully biological physics can offer some new tools, both experimental and theoretical, to advance research on the inner workings of the brain.

A clear understanding of the physical principles behind consciousness and intelligence could guide new treatments for injuries to the brain, mental illness (one of the largest emergent markets for present day pharmaceuticals, but with little underlying quantitative understanding) and to help the creation of new faster and smarter computers. The hope is that these smart computers will become effectively kinder and more perceptive than the average human being. These developments in turn would help revolutionise the modern world and thus enrich everyone's lives.

3.2 Suggested reading

Introductory

Gibb B J 2007 *The Rough Guide to the Brain* (London: Penguin). A compact introduction to the biology of brains.

Dehaene S 2014 *Consciousness and the Brain: Deciphering How the Brain Codes our Thoughts* (New York: Viking). Includes a popular discussion of MRI experiments in brain research performed at the Collège de France that provide both evidence for the flash light of consciousness and its treatment as a phase transition.

Sterling P and Laughlin S 2015 *Principles of Neural Design* (Cambridge, MA: MIT Press). A semi-popular account of quantitative neural science that explains some issues involved with the human connectome project.

Buzsaki G 2011 *Rhythms of the Brain* (Oxford: Oxford University Press). Fascinating coverage of electroencephalography (EEG) measurements of the electrical activity in brains which can be considered as a coherent binding phenomenon for mental processes.

Swanson L W 2012 *Brain Architecture: Understanding the Basic Plan* (Oxford: Oxford University Press). A compact account of brain anatomy.

Advanced

Dehaene S and Changeaux J P 2011 Experimental and theoretical approaches to conscious processing *Neuron* **70** 200. A review of modern tools to measure consciousness.

Chaikin P M and Lubensky T C 1995 *Principles of Condensed Matter Physics* (Cambridge: Cambridge University Press). Describes the modern theory of phase transitions applied to traditional areas of condensed matter physics.

2017 Allen brain atlas. (www.brain_map.org). Freeware software tools for visualisation of published brain data such as the Allen brain atlas.

3.3 Technical details

3.3.1 What is a phase transition?

Phase transitions play a central role in condensed matter physics and materials science. It has been known for many hundreds of years that the properties of materials sensitively depend not only on their chemistry, but also the phase adopted (solution, liquid, gas, etc), and these phases are determined by the physical environment, e.g. temperature, pressure, density, etc. Varying the physical environment of a material can cause sudden changes in its properties (a first-order phase transition, such as melting an ice cube) or a much more gradual change (a second-order continuous phase transition, such as ferromagnetism in a block of iron).

3.3.2 Order parameter

To parameterise a phase transition, an order parameter needs to be defined (Barrat and Hansen 2003). This needs to have a value of 1 in the high order phase and a value of 0 in the disordered phase (figure 3.1). If the parameter changes continuously at the point of the phase transition, it is defined as a continuous phase transition and it is a first-order phase transition otherwise. Famously, Kenneth G Wilson was awarded a Nobel prize for his theoretical contributions to understanding how phase transitions occur near to critical points and universal behaviour was found for them based on very general symmetry requirements. Similar ideas can be applied to biology for transitions in soft matter (e.g. helix/coil transitions or globule/coil transitions) and also for the neural activity of the brain.

3.3.3 BOLD contrast with magnetic resonance imaging

MRI is now used as a standard technique to diagnose patients in hospitals and it is based on the physical phenomenon of *nuclear magnetic resonance*. Nuclei in the sample absorb radio frequency photons at different rates and different frequencies dependent on their local atomic environment (they are thus sensitive to molecular bonding patterns) and the applied magnetic field. A gradient in the applied magnetic field means the resonant frequencies are shifted depending on the spatial position of the nuclei in the field and this forms the key process in MRI, i.e. it allows images to be formed.

BOLD (blood oxygen level dependent) describes a contrast mechanism that can be used in MRI experiments, in which the resonances are probed for oxygen atoms bound to haemoglobin in the blood (Buxton 2009). Specifically, with brain imaging, large changes in BOLD contrast correspond to the increased activity of the neurons in the regions imaged, i.e. they are metabolising oxygen at an increased rate.

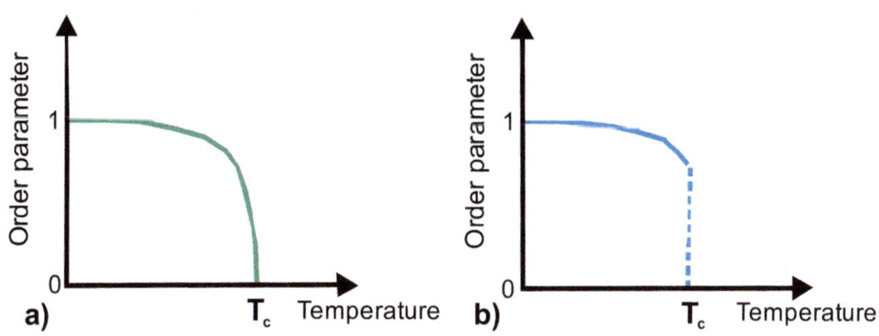

Figure 3.1. Order parameters for (a) a continuous phase transition (e.g. ferromagnetism or superconductivity) and (b) a first-order (e.g. solid/liquid) phase transition near to the critical temperature.

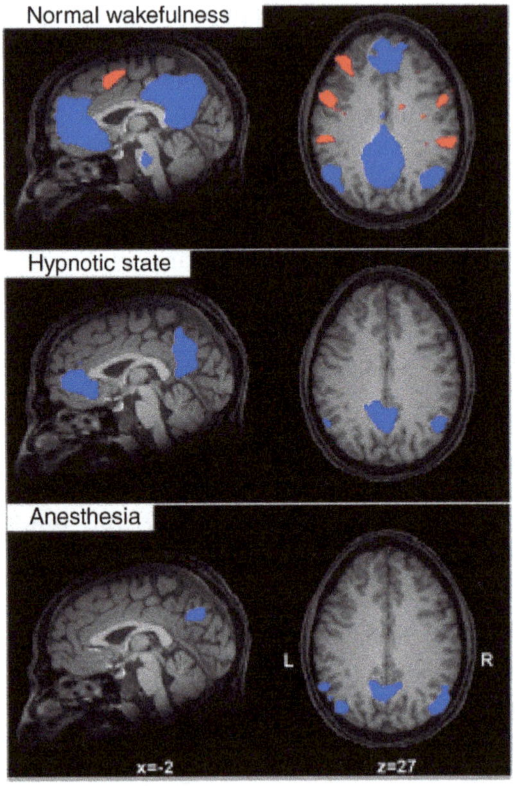

Figure 3.2. MRI scans with BOLD contrast show patterns of activity in a normal wakeful, a hypnotised and an anesthetised human brain. (Heine 2012).

Figure 3.2 shows some BOLD images of a human brain in normal wakefulness, hypnotised and anesthetised states (Heine 2012). Large changes are seen in the activity of different regions and BOLD MRI allows mental processes to be mapped with unprecedented detail. Nevertheless, the resolution is still fairly low at around 1 mm, many orders of magnitude away from that required to image single neuronal synapses and map the full neuronal network of an active human brain.

References

Barrat J L and Hansen J P 2003 *Basic Concepts for Simple and Complex Fluids* (Cambridge: Cambridge University Press)

Buxton R B 2009 *Introduction to Functional Magnetic Resonance Imaging: Principles and Techniques* (Cambridge: Cambridge University Press)

Heine L 2012 Resting state networks and conciousness *Front. Psychol.* **3** 1

IOP Publishing

Some Critical Questions in Biological Physics
A guided tour around the bugbears
Thomas Andrew Waigh

Chapter 4

Spike trains and the senses—the mouse's whiskers

4.1 Introduction

In addition to neuronal cells' roles in parts of the brain's inner workings, such as thought processes, the cells are also important for the brain's communication with the rest of the body. Information is transferred by means of electrical spikes along the body of neuronal cells and these spikes are then combined with the transport of small bags of neurotransmitters, which move between neighbouring cells (across the synaptic clefts). The bags bridge the transmission of information across the gaps between the cells. All the electrical spikes that travel in a particular neuron look very similar, so the information they transmit is immediately deduced to be due to variations in their timing. The bags of neurotransmitters also tend to act in a standardised fashion, although a variety of molecules can modulate their long term behaviour. This modulation phenomenon in turn affects the strength of synaptic connections and the nature of the information transferred between neurons. The synaptic connections form the molecular basis of memory and their modification determines the learning processes that enable memory storage in organisms over long time scales.

Memories form a central component of someone's personality. They allow people to perform their day-to-day tasks, such as locating their car keys, and help them to learn from their experiences, e.g. to avoid leaving their ballpoint pens in their trousers when they are placed in the washing machine. Good theories in biological physics exist for memories in the form of neural network models and the formation of memories is understood with great precision in simple organisms, e.g. Eric Kandel's Nobel prize winning work on sea slugs. The transmission of sensory information with spike trains is, however, slightly less well understood.

A good specific and more readily tractable question in brain research compared with consciousness or intelligence, thus considers how the spike trains that travel along neurons relate to the information transmitted by the senses that is subsequently

doi:10.1088/978-0-7503-1374-2ch4 © IOP Publishing Ltd 2017

interpreted by the brain. There are many fascinating problems still left in this field, but researchers have made better progress than that achieved with consciousness and intelligence, which gives more optimism for future advances. Simply put, researchers want to understand how information from the eyes, the ears, the nose, the skin, etc is encoded into neural activity and then subsequently decoded by an organism's brain. Such sensory responses frame people's view of the world and inform most of their behavioural decisions.

For humans, sensory information is very important for their quality of life. Even fairly minor perturbations, such as blocked ears or strained eyes, immediately lead to compromised lifestyles, and a very major reorganisation of people's lifestyles are necessitated by permanent impairment of a sensory faculty. There are thus many medical procedures to replace faulty sensors in the field of regenerative medicine that are caused by aging, disease and injury. To date the biggest success story has been the creation of cochlear implants that can return hearing to people who are completely deaf. Replacement lenses for cataract operations have also been extremely successful, but they do not directly affect any electrophysiological wiring, so are not considered further in this discussion. The effective use of cochlear implants requires that the necessary neural wiring to the brain is still intact and the brain is still able to make sense of all the acoustic information if it is resupplied with it (to some extent this ability to process information is a learned response in early life and can atrophy). Neural grafts in reconstructive surgery can also be very successful in the recovery of sensory faculties, e.g. for a patient to feel their toes after an operation, when the nerves have been severed in a road accident.

Beyond cochlear implants and neural grafts, there are many other large question marks hanging over the biophysics of the senses, and the classical biophysics approach is to consider the behaviour of simpler organisms to provide the opportunity for a clear-cut answer with fewer complications from behavioural responses. Rodents have an additional sense, not available to humans, that allows them to create three-dimensional images of their surroundings in complete darkness; they use their whiskers. Rodents are not unique in this respect, since some of their predators (cats) also have similar whisker based imaging capabilities. Do not try cutting off a cat's whiskers, because it results in a seriously peeved moggy with an impaired night imaging capability. The field has given rise to some of the most prosaic titles of research articles, such as *A night in the life of a rat: vibrissal mechanics and tactile exploration.*

As a specific example of whisker research consider the mouse. Every mouse has approximately 35 sensory whiskers on either side of its head, and each whisker has around 150 neurons that measure its response to the mouse's surroundings. Good quality data for the voltages in mouse neurons indicates that the mice are in part sensitive to the curvature of their whiskers, and thus the mechanical bending (the moment) of these hairs at their base where they meet the skin. There is also good anatomical evidence that shows that sensory neurons attached to the whiskers have widely varying architectures. Some of the neurons are wrapped around the hairs, some are to the side and others are at different vertical positions. It is expected they may constitute some kind of sophisticated strain gauge (e.g. a differential strain gauge to measure the deflections of the whisker), but the exact function of all these neurons is unknown.

L. *Question:* Provide a quantitative explanation of the role(s) of the one hundred and fifty sensory neurons that surround each whisker of a mouse.

In principle one or two neurons should be enough to measure the curvature of a single whisker. What are the additional 149 or 148 neurons doing? There appears to be an awfully large amount of redundancy in their activity, if the neurons are just needed to measure the curvature of a single whisker. Integration over many identical neurons could increase the signal-to-noise ratio and increase the accuracy of positional and temporal detection, and consequently improve whisker based imaging in the mouse brain. Nevertheless, it might be expected that all the neurons would be of a similar type and position, which is not the case. There are multiple distinct forms of neurons at multiple locations. How the mice integrate all the information together from all the different neuronal positions around a single whisker is not understood.

Simple mechanical models of a whisker that consider it to be a thin conical cylinder made from a uniform elastic material (technically it is called a cantilevered beam geometry) immediately encounter some challenges. The Euler–Bernoulli equation is required to describe the deformation of a whisker and it is a non-linear equation. Non-linear equations are tricky to solve and typically lack any analytic solution, i.e. nice compact algebraic solutions just do not exist. Experimentally the mice are placed among a field of vertical posts and a movie is made of the collisions of their whiskers with the posts to provide a well-defined sensory environment that acts as a standard reference system. Therefore, the Euler–Bernoulli equation needs to be solved with time to describe the interaction of a whisker with a post to calculate the resultant forces transmitted to the base of the whisker (and thus to the mouse's neurons). Mathematically it is a tractable (doable) problem, but technically it is a little challenging, and only a few specific numerical solutions are currently available. Experimentally it would also be useful to attach force transducers (micro-Newton force sensitivity is required, since whisker forces are minute, but this sensitivity is readily available with modern piezo transducers) to the posts to further constrain the models, e.g. it could allow for the simultaneous determination of the elastic modulus of the whisker and the force on the whisker's base to provide a more accurate picture of whisker based sensing.

L. *Question:* Solve the Euler–Bernoulli equation with time for a conical whisker in contact with a post in three dimensions and use it to describe some mouse experiments.

The next challenge is to correlate the electrical activity of the neurons with the forces, torques and moments experienced by the base of each whisker to which they are attached. Currently the most successful model assumes the frequency of the neuronal spike code that is transmitted to the mouse's brain is directly proportional to the change in the curvature of the whisker at its base. Although this is a good starting point, it is expected that a great deal of additional information will also be transmitted to the brain from the responses of each individual whisker, e.g. the angle of elevation of objects in the surrounding environment. More accurate experiments will drive the creation of better models and the process can be iterated to provide a more detailed picture of whisker based imaging.

Additional complications are associated with the transmission of information from sensory organs to the brain that involves the activity of neurons. Compression of sensory signals is often required, because all naturally occurring neurons have a fairly limited band width (the range of frequencies over which they can transmit information), and important sensory information is distributed over a wide range of values that can vary over many orders of magnitude. In general, this leads to a logarithmic law of sensitivity (Weber's Law) for a large number of senses, such as eyesight, hearing, touch, smell, taste, etc. For instance, this is why sound is measured in decibels; a logarithmic unit that is matched to the sensitivity of human hearing, which in turn is limited by the bandwidth of the auditory neurons, i.e. it may be the evolutionary response to this constraint. In addition to compression, normally a first step in biological signal transmission is for redundant information to be subtracted off signals to optimise the amount of useful information that is carried, e.g. for mice only changes in whisker curvature need to be transmitted, since the absolute values of the unperturbed whisker forces do not contain any useful information, i.e. there is often some curvature in undeformed whiskers that should be neglected for efficient transmission of information. The contrast of signals can then be enhanced to magnify their variations and the signals are then compressed for transmission to the brain.

An analogy can be made with fast broadband internet for people's homes. High information transfer rates mean high definition movies can be streamed on demand. WiFi has a more limited bandwidth than a physical broadband connection, which can lead to glitches in real time video streaming, such as problems with Skype during video conferencing. The senses are constrained by similar issues associated with bandwidth. Lots of data is sacrificed in human senses, to allow real time transmission and processing in the brain. The data sacrificed from measurements made by the eye is often the source of optical illusions experienced by humans, since it can lead to ambiguities in the interpretation of an image that were not there in the original signal that the retina detected, e.g. Mach bands.

One curious aspect of sensory biophysics is that of *hyperacuity* of the senses. Many senses outperform the accuracy expected based on simple physical calculations and they do this by using sophisticated neuronal computational tools. For example, bats can measure sub-millisecond signals during ultrasound imaging, a factor of a hundred times faster than expected from the maximum update speed of their individual neurons. They do this with *coincidence detection*; a spike is only emitted by a neuron if it receives two separate spikes within a sub-millisecond time delay. A similar mechanism was discovered by particle physicists in the 1920s and resulted in a Nobel Prize for Walther Bothe. Coincidence detection is still in use in positron emission tomography (PET) scanners, where the annihilation of a positron by an electron results in two photons that travel in opposite directions and the event can be localized in three-dimensional space through the use of a circular array of detectors around the sample, which time the arrival of the pair of photons. PET scanners are used in hospitals for medical imaging.

Another example of hyperacuity, this time with humans, is the angular resolution of the eye, which is much higher for extended objects than that expected from the diffraction limit of single rods or cones (the light sensitive cells at the back of the eye). Here the trick is again due to neuronal software, which in this case spatially integrates over many measurements to improve the signal accuracy. A similar type of trick is used in the colour vision of the human eye to create resolutions of the wavelength detected in coloured images with an accuracy of five nanometres from only three cone detectors that are sensitive to a very broad range of red, blue and green photons. A small proportion of the female population (around 2%) are now thought to have a fourth type of colour sensitive cone. However, this still does not explain the high degrees of wavelength sensitivity observed in humans and the answer is found in the sophisticated neural processing that is performed.

It is too soon to speculate as to whether mice achieve hyperacuity from neuronal measurements with their whiskers, but it seems to be a distinct possibility. Angstrom displacements of hairs (used as sound transducers) are detected in the ears of mammals, although the resolution of mouse whiskers probably falls far short of this extreme case of sensitivity.

One apparent anomaly in neuronal coding appears to have been recently explained. The basic premise in neuronal coding is that the frequency of the spike trains is proportional to the strength of the signal sensed. Nevertheless, reaction times for many organisms are often faster than the average time between two spikes, which makes it impossible to define a frequency in any tangible manner. The new explanation is *rank coding*, where simultaneous parallel measurements of an array of spike trains allows an organism to both react after the average duration of one spike and calculate the strength of the signal (based on its frequency) simultaneously. Sophisticated neuronal software allows the brain to do the necessary analysis of the spike statistics, i.e. the brain ranks the arrival times of the spikes. The rank coding mechanism also provides an alternative explanation for the multiple neurons found with each mouse whisker (i.e. they provide fast reactions with high levels of signal strength discrimination), but not the large variety of different types of neuron that occur.

The future of sensory biophysics is expected to be strongly associated with regenerative medicine. Cochlear implants are common medical procedures, but replacement electronic retinas (based on CCD cameras) have also now become a clinical reality. Nevertheless, much more fundamental biophysics research is needed to underpin these developments, to improve the success of these medical procedures and restore the quality of as many people's lives as possible.

4.2 Suggested reading

Introductory

Bossomaier T 2012 *Introduction to the Senses: From Biology to Computer Science* (Cambridge: Cambridge University Press). The connection between electrophysiological spikes and the senses is discussed from the perspective of computer programmers involved in the design of new computer games.

Advanced

Hartmann M J Z 2011 A night in the life of a rat: vibrissal mechanics and tactile exploration *Ann. N. Y. Acad. Sci.* **1225** 110–8. Introduces research in the field of whisker based imaging.

Campagner D, Evans M H, Bale M R, Erskine A and Petersen R S 2016 Prediction of primary somatosensory neuron activity during active tactile exploration *eLife* **5** e10696. Primary research that describes experiments with mouse whiskers, that connects whisker mechanics to the electrophysiological responses.

Rieke F, Warland D, de Ruyter van Steveninck R and Bialek W 1997 *Spikes: Exploring the Neural Code* (Cambridge, MA: MIT Press). A classic advanced textbook on neural codes, which contains a thought provoking discussion of information theory and hyperacuity.

4.3 Technical details

The neurons around the whisker in the snout of a mouse are arranged in a fairly complicated manner (figure 4.1). Exactly how all the neurons work together to

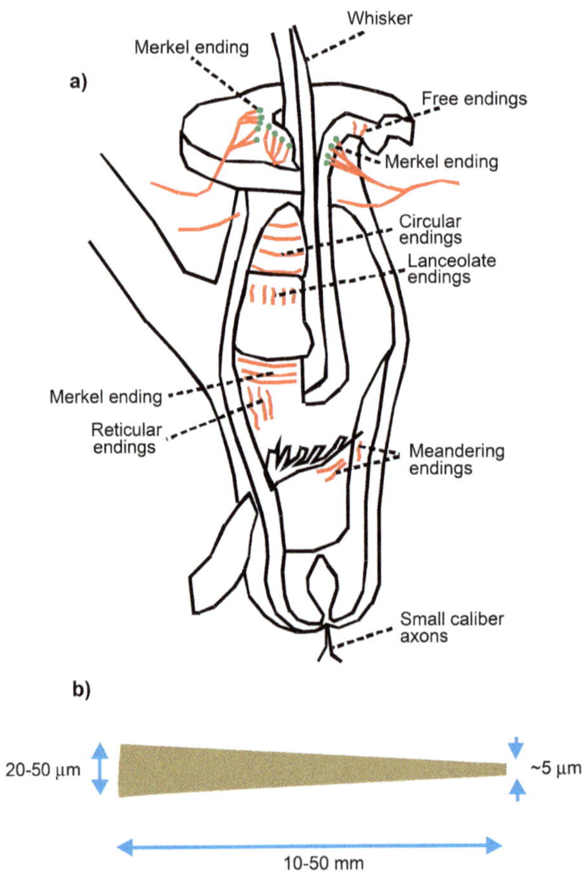

Figure 4.1. (a) Anatomical picture of the internal structure of a whisker in the snout of a mouse highlighting some of the nerve endings (red). (b) Schematic diagram of the dimensions of a mouse whisker.

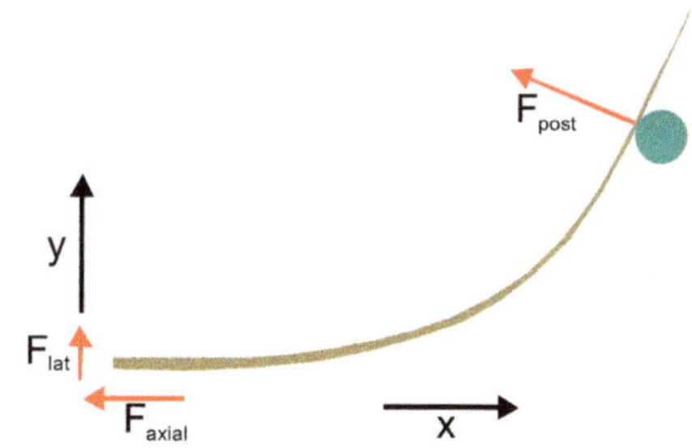

Figure 4.2. The directions of the lateral (F_{lat}) and axial (F_{ax}) forces at the base of the whisker, where it is detected by neurons, when a force (F_{post}) is applied by a post (green) on the whisker.

provide the mouse with sensory information is not well understood, but they are thought to act like a sophisticated strain gauge or an array of strain gauges.

4.3.1 The Euler–Bernoulli equation

The Euler–Bernoulli equation can be used to describe the shape of a whisker in a two-dimensional plane when a force is applied to the whisker (Lucianna *et al* 2016), e.g. a contact force with a pole (figure 4.2),

$$\kappa(x) = \frac{\mathrm{d}^2y/\mathrm{d}x^2}{\left[1 + \left(\frac{\mathrm{d}y}{\mathrm{d}x}\right)^2\right]^{3/2}} = \frac{M(x)}{EI} \tag{4.1}$$

where x is the axis initially along the undeformed whisker, y is the axis perpendicular to x, $\kappa(x)$ is the x dependent curvature of the whisker, $M(x)$ is the moment at each point along the whisker, E is the Young's modulus of the whisker and I is the area moment of the whisker.

For small deflections $(\frac{\mathrm{d}y}{\mathrm{d}x})^2$ is negligible in equation (4.1) and the equation can be linearized to give

$$\kappa(x) = \frac{\mathrm{d}^2y}{\mathrm{d}x^2} = \frac{M(x)}{EI} \tag{4.2}$$

This can be easily solved by integrating it twice to give a quadratic equation,

$$y = \frac{M(x)x^2}{EI} \tag{4.3}$$

Figure 4.3 shows this quadratic solution scaled with constants chosen so that it is relevant to the forces experienced by a mouse whisker (Campagner *et al* 2016).

In reality mouse whiskers are conical (figure 4.1(b)), which needs to be included in the model, since the tip of the whiskers deform much more easily than their base (Birdwell *et al* 2007). Furthermore, the whiskers often have a natural amount of curvature in their undeformed state. This problem has been a fourth year undergraduate project over the last two years in Manchester and the students have made

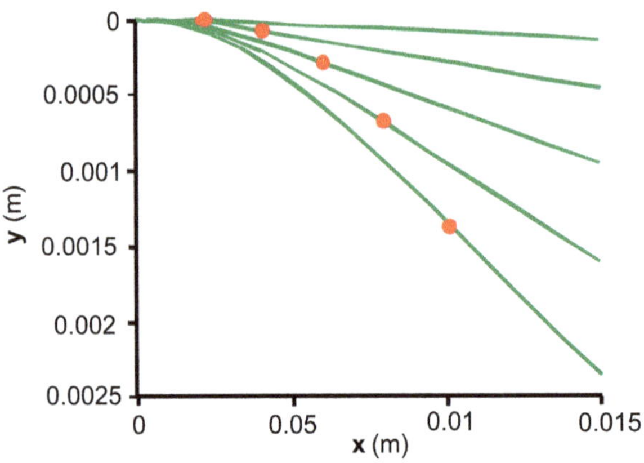

Figure 4.3. The deformation of a cylindrical whisker under a force of 7 mN applied in the negative y direction. Red dots represent the point at which the force due to the pole is applied (Courtesy of Ali Fensome and Tom Flavell).

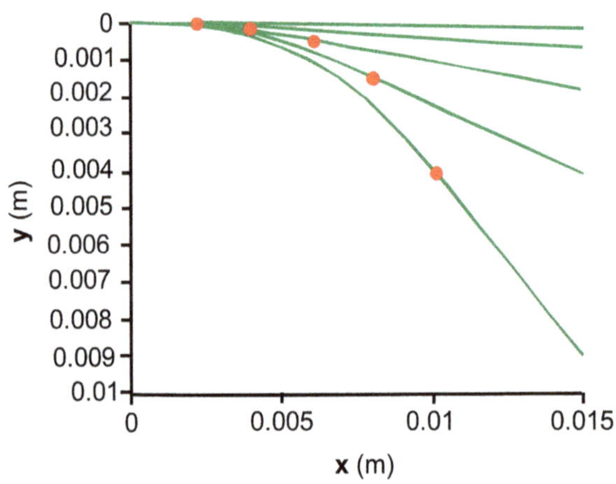

Figure 4.4. The deformation of a conical whisker under a force of 7 mN applied in the negative y directions. Red dots represent the point at which the force due to the pole is applied (Fertis 2006) (Courtesy of Ali Fensome and Tom Flavell).

some good progress with it. Figure 4.4 shows numerical solutions of the Euler–Bernoulli equations for a conical whisker with a linear taper, which again has curvature in a single two-dimensional plane (Fertis 2006). The degree of deflection depends sensitively on the position at which the force is applied (compare with figure 4.3; the deformations of the end of the tip are 4–5 times larger).

Whisking is the active oscillatory process that a mouse uses to move its whiskers around to scan its surroundings and it is driven by muscles in the mouse's skin. An additional fun experiment is therefore to explore the non-linear physics of phase portraits for the dynamics of whiskers, e.g. to plot the Euler angles (θ, φ and χ) of the whisker during the oscillatory whisking process, such as θ versus φ. Good models for these non-linear dynamics are still at a very early stage of development, but they

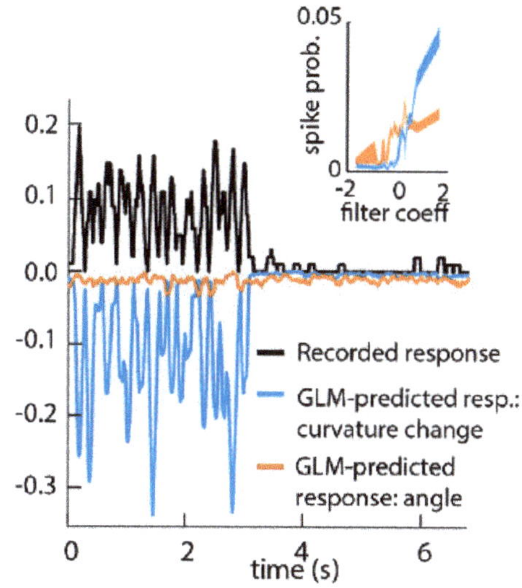

Figure 4.5. The measured electrical activity from a mouse's whisker (black) versus the predicted values from optical measurements of the curvature (blue) (from Campagner *et al* 2016).

could be used to inspire artificial designs, e.g. to provide sensory systems for robots to be used in mine detection.

4.3.2 Spike coding

A reasonably good first approximation to the electrical activity of neurons attached to mouse whiskers can be made based on a model that the neurons are only sensitive to the whisker curvature (figure 4.5). The model for the electrical activity does require some questionable assumptions, such as the linearity of the mapping between neuronal spikes and the amplitude of sensory stimuli (Eliasmith and Anderson 2003), but at least the theoretical model is reasonably well established. The current challenge is to connect the electrical activity with the response of the whisker, beyond its simple two-dimensional curvature, and do so in a more quantitative fashion.

References

Birdwell J A, Solomon J H, Thajchayapong M, Taylor M A, Cheely M, Towal R B, Conradt J and Hartmann M J 2007 Biomechanical models for radial distance determination by the rat vibrissal system *J. Neurophysiol.* **98** 2439–55

Campagner D, Evans M H, Bale M R, Erskine A and Petersen R S 2016 Prediction of primary somatosensory neuron activity during active tactile exploration *eLife* **5** e10696

Eliasmith C and Anderson C H 2003 *Neural Engineering* (Cambridge, MA: MIT Press).

Fertis D G 2006 *Non-linear Structural Engineering* (Berlin: Springer).

Lucianna F A, Albarracin A L, Vrech S M, Farfan F D and Felice C J 2016 The mathematical whisker: a review of numerical models of the rat's vibrissa biomechanics *J. Biomech.* **49** 2007–14

IOP Publishing

Some Critical Questions in Biological Physics
A guided tour around the bugbears
Thomas Andrew Waigh

Chapter 5

Elastic turbulence—gloopy chaos

5.1 Introduction

The living state of matter in organisms is often identified with motion, whether it be on a small length scale (cells or molecules) or on large length scales (whole organisms). To understand this motion an initial simplification is to consider the response of materials between two moving boundaries, e.g. sections of cartilage in a joint separated by synovial fluid. The material is considered to be passive and just reacts to the boundary motion, as opposed to active materials (chapter 2) that can drive the motion of fluids internally. Here some questions in the motion of passive materials will first be considered.

Osbourne Reynolds was one of the founding fathers of the University of Manchester and he also was a leading figure in the history of turbulence, so turbulence has had a long tradition at the university. Reynolds performed a classic series of experiments using long pipes through which liquid water flowed, with a dye to visualize the patterns of flow. Flows were found to become chaotic and unsteady at high flow rates, and the flow rate at which the transition occurred was repeatable and depended only on the viscosity of the fluid, the density of the fluid and the width of the pipe. As such, the transition to turbulent flow is seen to be a universal phenomenon, relatively insensitive to other details of the flow geometry. A simple mathematical description was then introduced by Reynolds to predict when all types of flow would become turbulent, i.e. when flows become unsteady with lots of randomly fluctuating vortices (water flowing down a plug hole type motions). This turbulent flow is an extremely common occurrence in everyday life, seen in smoke plumes from chimney stacks, travel on airplanes (when the seatbelt warning light goes on), and the design of swim suits that emulate the patterns found on shark skin. The original mantra based on Reynolds research, which is repeated in most fluid mechanics textbooks, is that low Reynolds number fluids do not demonstrate turbulence. This means that flows with low velocities, with low density fluids, at small length scales (e.g. in small blood vessels or with the swimming patterns of

 © IOP Publishing Ltd 2017

microorganisms) or with high viscosities should all be stable (the jargon is *laminar*) without any turbulence.

However, there have been many theoretical developments, since the time of Reynolds in the 1800s, and questions in the field of flow instabilities are now posed under much more general conditions. Any non-linear terms in the equations that describe the motion of viscoelastic fluids need to be defined to determine the existence of unstable flows. It is these non-linearities that give rise to the phenomenon people interpret as turbulence; the chaotic motion that drives vortices through a fluid.

Surprisingly, recent experiments with complex fluids demonstrate that exceptions to the Reynolds theory exist. Take a concentrated (jelly-like in consistency) solution of long DNA chains and force it to flow between two plates. Technically the concentrated DNA solution is not a gel, because there are no cross-links between the polymer chains, but it is a very viscous gloopy fluid. Above a critical flow rate, the DNA chains' motion becomes unsteady and erratic, similar to the case observed with normal turbulence. However, the Reynold's number can be as low as 10^{-5}, which is one hundred million times smaller than the critical value expected from the Reynolds theory (the threshold value measured by Reynolds for his characteristic number was 5000). Reynolds theory has completely failed!

In this phenomenon of elastic turbulence, separate equations that describe the elasticity are thought to give rise to the non-linearities that cause turbulence, not the Navier–Stokes equations that describe the motion of purely viscous solutions that Reynolds based his research on. How the non-linearities do this in detail is not very well understood, and it currently appears to be fairly system dependent. One possibility with the DNA solutions on which the Manchester laboratory has recently performed experiments, is that it is due to the coil-stretch transition of the individual DNA chains, and the degree of chain stretching becomes a strongly non-linear function of flow rate in rapid flows. Random extension and then relaxation of the chains may give rise to the rapid fluctuations of the velocities seen in elastic turbulence with DNA.

In terms of a general picture, there are either lots of varieties of elastic turbulence that can occur, or the key control parameters have not yet been identified, that reduce the diverse range of data sets measured on the phenomena down to a single fundamental master curve that would then hint that a universal explanation is possible. Conceivably the elastic turbulence effect is also closely related to surface interactions, which cause slippage of the elastic fluids on the boundaries of the flow geometry. Elastic turbulence is also thought to be associated with the phenomenon of normal stresses observed in the *Weissenberg effect*. If a rotating rod is placed in a viscoelastic solution which has an unusually large normal stress coefficient, the fluid can climb up the rod, and the effect is named after Weissenberg. The *normal stress* in complex fluids couples a stress in one direction to that in a perpendicular direction. The Weissenberg experiment can be done in one's own kitchen with a food blender and some pancake batter.

A slightly surprising area in which elastic turbulence is also thought to play a role is that of shampoo. The current holder of the Isaac Newton Lucasian chair at the

University of Cambridge is famous for his theoretical models of shampoo (worm-like micelles), among other subjects. Soap molecules in shampoo can self-assemble into giant worm-like aggregates (micelles). These aggregates provide the soap solutions with substantial amounts of elasticity, which allows them to be conveniently applied to the hair. Hand in hand with the increased elasticity provided by the worm-like micelles, comes the phenomenon of elastic turbulence, which has major implications for the processing conditions required inside soap factories, e.g. the materials move like a plug as they flow down pipes (rather than the parabolic profiles for conventional fluids that follow the predictions of Newton) and unusual flow instabilities are seen when the pipes change direction.

Another even more extreme example of a non-linear flow phenomenon can also occur in shampoo, which is *shear banding*. Stripes spontaneously form in shampoo solutions when they are sheared, in which the stripes contain different concentrations of soap, and flow at different velocities. Such behaviour is not well described by the standard Newtonian prescription in which a constant velocity gradient is expected. There should be no heterogeneities in the flow behaviour of shampoo according to Newton. Shear banding is also found in biological samples, such as solutions of DNA and cell scaffolding proteins (actins), although whether it is the same phenomenon as observed with the soap solutions is open to debate.

Applications for elastic turbulence in medicine concern flow phenomena in the lungs of diseased individuals. The internal surfaces of healthy lungs are covered with a thin layer of a runny viscoelastic fluid; mucus. In diseased states the mucus can become too thick and elastic to be efficiently transported by the tiny hairs (cilia) found on the lungs, it irreparably damages the cilia, and the patient has to continually cough to clear their airways. Elastic turbulence may therefore play a role in *cough*; an awkward (often painful) mechanism used to mobilise the elastic aggregates that block people's airways.

A number of other applications of elastic turbulence have also been recently discovered with starches during food processing. Starch represents mankind's principle food stuff. The flow behaviour (rheology) of starch is thus important for the manufacture of a huge range of food products, e.g. potato starch extruded into Monster Munch shapes for crisps or wheat starch formed into spaghetti. During processing starch has been shown to demonstrate elastic turbulence, which can complicate the industrial conditions used, and leads to wastage of energy and materials.

Polymers are also often injected down oil wells to improve extraction procedures in the petroleum industry and they are known to demonstrate elastic turbulence when confined to porous rocks during these recovery procedures. How this affects oil reclamation in detail is not well understood.

Another application is seen when people attempt to miniaturise devices in biotechnology using microfluidics, such as those needed for high throughput DNA technologies. A problem with such miniaturised devices is that they have low Reynolds number flows that do not mix well, since there is no classical turbulence. The need for uniformly mixed reactants is a common prerequisite for the success of the vast majority of chemical reactions. For example, if DNA needs to be manipulated with restriction enzymes the two components need to be properly mixed together. Elastic turbulence provides an alternative mechanism that can very effectively mix reactants in small capillaries at micron length scales in the microfluidic devices.

Due to its importance in so many everyday phenomena, elastic turbulence clearly deserves a better understanding.

L. *Question*: Create an accurate quantitative model for elastic turbulence in a complex fluid.

A cynic's response to this question might be that the standard vanilla type of classical turbulence has evaded a fully quantitative description over the last 100 years (famously it was called the most important unsolved problem in classical physics by Richard Feynmann, who died in 1988, and still no-one has provided a major breakthrough in our understanding), so why should better success be expected with an exotic tutti-frutti flavour, elastic turbulence? Furthermore, a general solution to Navier–Stokes equation is still one of the top 10 unsolved mathematics problems in the 21st century. The solution of this equation is intimately connected with the prospects of a quantitative theory for turbulence. A pragmatic response to such extreme challenges is to perhaps set our sights slightly lower. Maybe some modest contributions can be provided to this field that will inevitably fall short of a fully predictive theory, but will give some important new insights to a field that is both technologically and medically important.

It is possible to break down the problem of elastic turbulence into sub-problems. First of all a better understanding is required of how the velocity varies from point to point in a viscoelastic sample that experiences elastic turbulence when it flows.

L. *Question*: Measure the length dependence of the velocity fluctuations in a material that experiences elastic turbulence.

There are a number of ways to perform these experiments using velocimetry techniques, such as particle tracking or optical coherence tomography. A model is then required to describe the behaviour observed and here is where the main challenges lie. A scaling approach would be a first good place to start. Scaling laws are a standard method to create a broad-brush stroke solution to a physical problem, before a fully quantitative model can be created. Richardson (1922) and Kolmogorov (1941) created a famous scaling approach for classical turbulence, which is reasonably successful and currently it is the main paradigm for understanding velocity statistics in turbulent fluids. A wide range of sizes of vortices rotate in turbulent fluids and the larger vortices dump energy into the viscosity of the smaller vortices in a self-similar manner. An analogous scaling approach relevant to viscoelastic flows is required to describe elastic turbulence, although the underlying physical factors driving the motion may be completely different.

The velocity fluctuations in elastic turbulence are intermittent; they start and stop erratically, but is it possible be a little more specific about how these fluctuations occur?

L. *Question:* Calculate the time dependence of the velocity fluctuations in a material that experiences elastic turbulence.

The spectrum of the velocity fluctuations thus needs to be carefully measured in experiments with high resolution velocimetry methods and the Kolmogorov model for classical turbulence then adapted to describe it. Preliminary experiments show there is a sharp change in the form of a power law decay of the velocity spectrum (the power spectrum of the velocity as a function of frequency) as the shear rate is increased and a sample is driven into the elastic turbulence state, where the power law exponent of the velocity spectrum on the frequency changes from around 1.5 (at low shear rates) to 3.5 (at high shear rates). Nevertheless, the fundamental origin of this behaviour is still unknown. There is already a possible explanation for the velocity spectrum in elastic turbulence with dilute non-interacting polymers that exists in the literature. Unfortunately, the effect also occurs in concentrated solutions of polymers with very strong inter-particle interactions, invalidating the theory in these cases, so no general theory yet exists. An additional complication is the anti-turbulence effect that occurs when dilute concentrations of polymers are added to Newtonian fluids. Add dilute polymers to the water in a fireman's hose and the water can travel three times further, since less energy is dissipated in turbulence. Thus, the addition of polymers first damps classical turbulence and then causes a new type of elastic turbulence as the concentration of the polymers increases.

In classical turbulence the flow near to the surfaces of bodies is perturbed in a well-defined manner; there is a *boundary layer*. Such boundary layers are crucial in the design of high speed vehicles, such as Formula One cars and fighter aircraft, where the control of turbulent processes is paramount for their performance. Quantitative analysis of boundary layers has led to new inventions, such as ultra-fast swim suits that mimic the wavy patterns on shark skins to reduce the energy dissipated in turbulence created during swimming. The suits were banned from competitions in 2009, since they provided swimmers with too large an advantage (17 world records were broken at a single meeting in Croatia in 2008 using the suits). It is thus natural to question whether a boundary layer also occurs in elastic turbulence.

L. *Question:* Quantitatively describe the nature of the boundary layer in a material that experiences elastic turbulence.

A few preliminary experiments already demonstrate evidence for such a boundary layer in concentrated solutions of polymers with very long chain lengths, but many more studies are needed to satisfactorily define all the quantitative phenomena involved.

Other exotic forms of turbulence exist in addition to elastic turbulence, such as bacterial turbulence (chapter 10) and liquid crystalline turbulence (turbulence in liquid crystalline materials, e.g. composed of rod-like molecules). Bacterial turbulence has implications for the study of infection, water treatment and the wide range of uses of bacteria in biotechnology to produce biomaterials, e.g. genetically modified organisms that produce biodegradable plastics. The most common application for liquid crystals is their use in display screens for electronic devices. Issues with turbulence may be important for the processing conditions required to make such liquid crystalline devices. Clearly phenomena in bacterial turbulence and liquid crystalline turbulence also require more detailed quantitative studies.

In the short term, the development of new instruments to study microfluidic flows is expected to be very useful for the study of elastic turbulence. Some of the issues involved in this field of *microrheology* are considered further in chapter 13; the study of flow behaviours of viscoelastic samples at small length scales. Hopefully a better understanding of elastic turbulence will be available in the near future and it will help guide the design of new chemical factories, treatments for lung disease, procedures in oil recovery and microfluidic mixing devices used in biotechnology, amongst many other applications.

5.2 Suggested reading

Introductory

Piazza R 2011 *Soft matter: the stuff that dreams are made of* (Berlin: Springer). Provides a general introduction of the flow properties of soft fluids.

Goodwin J W 2000 *Rheology for chemists: an introduction* (London: RSC). Describes some of the simple experimental and theoretical tools used in rheology to understand the flow behaviour of materials.

Advanced

Groisman A and Steinberg V 2004 Elastic turbulence in curvilinear flows of polymer solutions *New J. Phys.* **6** 1–47. Primary research literature that describes the strange elastic turbulence phenomena found in solutions of long polymeric molecules (specifically polyacrylamide) from the perspective of classical turbulence theory.

Frisch U 1995 *Turbulence: The Legacy of A.N.Kolmogorov* (Cambridge: Cambridge University Press). Elegant compact introduction to the mathematics that describes the statistics of classical turbulence.

Jun Y and Steinberg V 2011 Elastic turbulence in a curvilinear channel flow *Phys. Rev.* E **84** 56325. Evidence for a boundary layer in the flow properties of a polymeric solution that experiences elastic turbulence in a microfluidic channel.

Morrison F A 2001 *Understanding Rheology* (Oxford: Oxford University Press). Introduces some of the mathematical techniques needed to understand the unusual flow behaviour observed in complex fluids, e.g. normal stresses.

Fouxon A and Lebedev V 2003 Spectra of turbulence in dilute polymer solutions *Phys. Fluids* **15** 2060–72. Primary theoretical research literature that describes a model for the velocity fluctuation spectra in elastic turbulence.

Dean B and Bhushan B 2010 Sharkskin surfaces for fluid drag reduction in turbulent flow: a review *Phil. Trans. R. Soc.* A **368** 4775–806. This review considers the impact of boundary roughness on classical turbulence. Riblet patterns found on shark skins (e.g. fast swimming Marko sharks) cause a substantial reduction in the energy dissipated during turbulence. The impact of exuded mucin on classical turbulence with swimming sharks is also discussed, i.e. viscoelastic boundary layers.

5.3 Technical details

5.3.1 Turbulence

Turbulence is an infamously tricky problem and even some very heavy duty mathematics techniques have thus far failed to fully describe the phenomenon (Davidson 2015). Useful models exist for very specific phenomena in turbulence, such as the nature of boundary turbulence, but these descriptions are fragmentary and do not provide a satisfactory general description. Navier–Stokes' famous equation for the dynamics of any classical fluid is

$$\frac{\partial v}{\partial t} + (v.\ \nabla)v = -\frac{1}{\rho}\nabla p + \eta\nabla^2 v \tag{5.1}$$

where v is the fluid velocity, ρ is the density, t is time, p is the pressure and η is the viscosity. The equation is the basis of most standard approaches to fluid mechanics and thus also for turbulence. Equation (5.1) is non-linear. It is tricky to solve and the mathematicians have not yet found a general solution to it (neither can they say categorically whether one exists).

5.3.2 Kolmogorov/Richardson theory for classical turbulence

Classical turbulence is characterised by the intermittency of the velocity fluctuations, i.e. the velocity at a particular point in the fluid starts and stops in an unusual

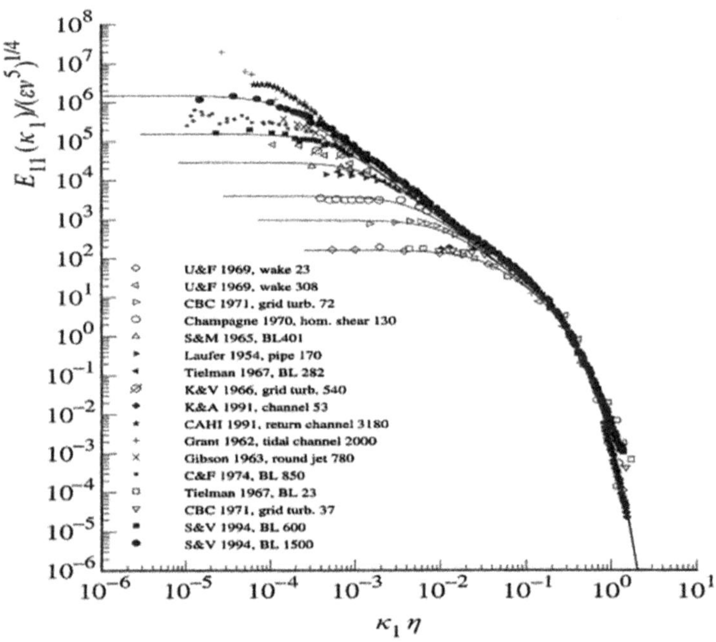

Figure 5.1. The energy dissipated ($E_{||}$ related to the velocity through $E_{||} = \frac{1}{2}mv^2$) as a function of length scale (η) in classical turbulence based on a wide range of different experiments from different laboratories, i.e. it is a universal phenomenon (Davidson 2015, Saddough and Veeravalli 1994. © Cambridge University Press. Reprinted with permission).

manner with characteristic statistics. The Kolmogorov/Richardson theory is perhaps the most successful model to describe the statistics of classical turbulence and it is based on a simple robust scaling derivation. Specifically, the model considers an energy cascade of eddies from large length scales to smaller length scales where the energy is finally dumped into the viscosity of the fluid as heat. The moving components are modelled with different sized vortices (Guyon *et al* 2015).

5.3.3 Turbulent spectra

In classical turbulence the spectrum of the velocity fluctuations in fully developed (well above the transition point) isotropic (sufficiently distant from the flow boundaries) systems has a robust repeatable characteristic form. A wide range of data from different groups can therefore be superposed on a single master curve (figure 5.1). The scaling behaviour of the master curve is in reasonably good agreement with the Kolmogorov/Richardson theory, but the predictions are not fully quantitative and only describe broad trends in the data (Frisch 1995).

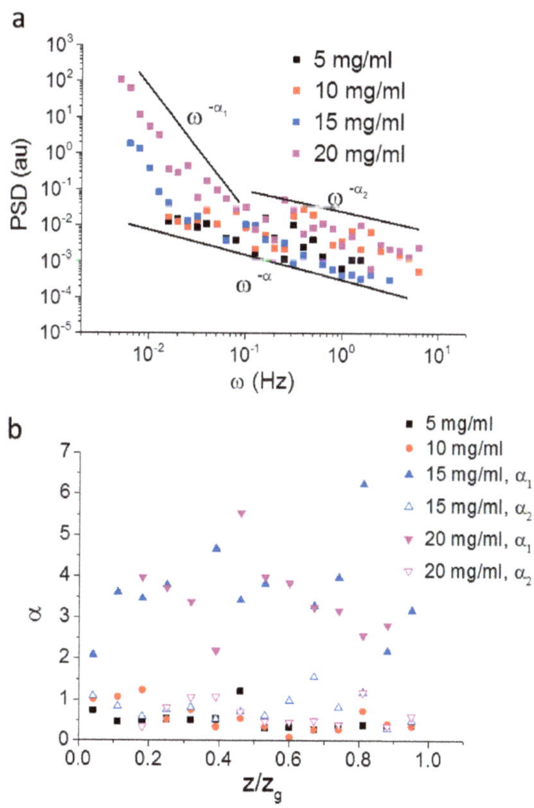

Figure 5.2. Elastic turbulence in concentrated DNA solutions at different DNA concentrations. (a) Shows the power spectral density (PSD) of the solution velocity as a function of frequency (ω). (b) The exponent of the power law decay of the PSD as a function of position in the plate/plate geometry (z/z_g) (Malm and Waigh 2017). Reprinted by permission from Macmillan Publishers Ltd. The article is available under a CC-BY-4.0 licence.

In elastic turbulence our picture of the spectrum of velocity fluctuations (very closely related to the energy spectrum in figure 5.1, since the kinetic energy is related to the velocity) that occur is still fragmentary. Experiments in our own laboratory using optical coherence tomography velocimetry (chapter 8) indicate that a different characteristic spectrum occurs for the flow of DNA chains compared with classical turbulence (figure 5.2), following the scaling PSD$\sim\omega^{\alpha}$, where α is approximately 3.5 and the Kolmogorov prediction is 1.67. This type of large exponent was found previously in a study of Groisman *et al*, although whether these results are directly comparable is open for debate (Groisman and Steinberg 2000). The Groisman study claimed to be in the dilute concentration regime for synthetic polyacrylamide, whereas we were studying much more concentrated specimens of DNA (Malm 2015, Malm and Waigh 2017).

Much more research is required in this field to satisfactorily answer all the outstanding questions. However, with DNA, there clearly is a non-linear flow instability at high shear rates, which is shown by the velocity colour plots in figure 5.3(d) (an unstable intermittent velocity profile is seen associated with elastic turbulence).

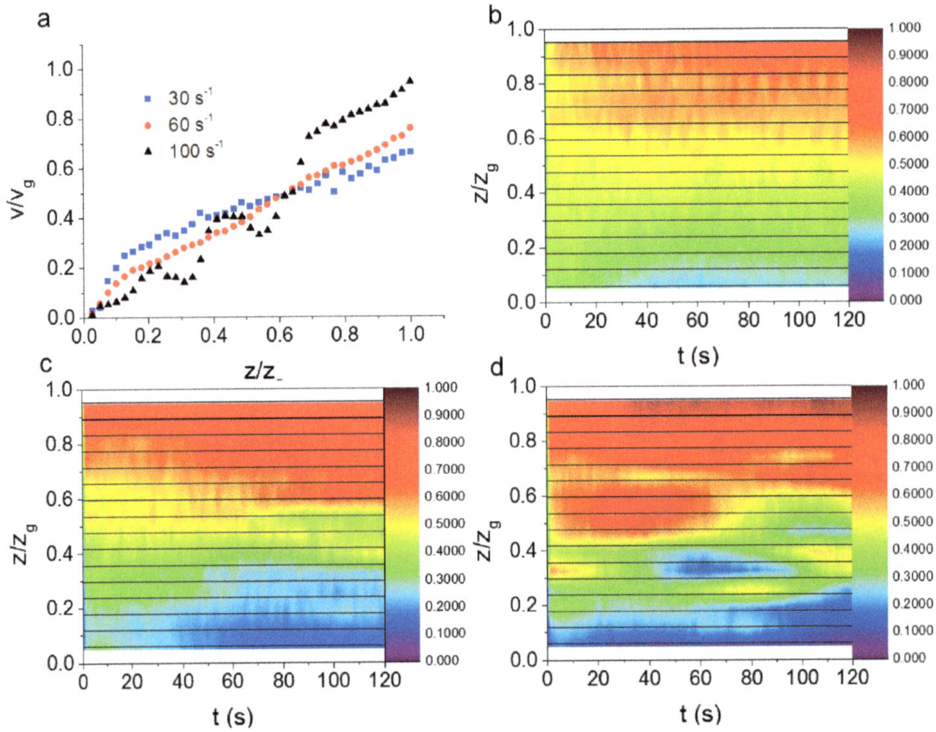

Figure 5.3. (a) Normalised shear velocity (v/v_g) as a function of normalised distance across the rheometer gap (z/z_g) for a 15 mg/ml solution of DNA (high salt) at three different steady shear rates that show significant wall slip and irregular flow when sheared at 100 s^{-1}. The colour maps show the normalised shear velocity as a function of the normalised distance across the rheometer gap for the same sample and the time after shear start-up at constant shear rates of b) 50 s^{-1}, c) 60 s^{-1} and d) 100 s^{-1}. The horizontal lines show the depths at which measurements were made, and the color scale is normalised to the top plate velocity (Malm and Waigh 2017). Reprinted by permission from Macmillan Publishers Ltd. The article is available under a CC-BY-4.0 licence.

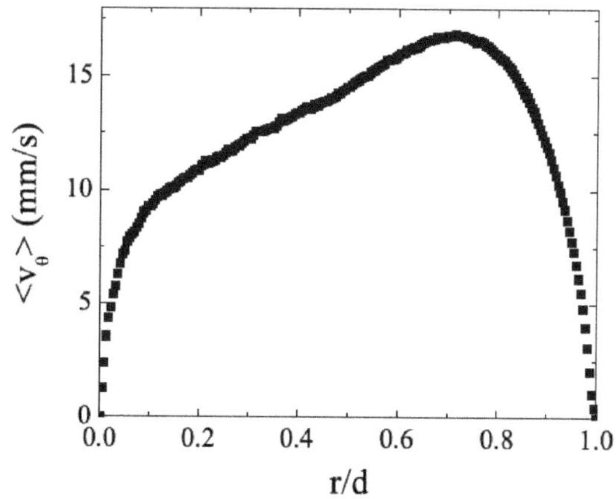

Figure 5.4. The mean longitudinal velocity profile as a function of position (r) across a channel that contains a solution of polyacrylamide that flows with a Weissenberg number of 951. A significant boundary region can be observed on both the upper and the lower walls. Reprinted figure with permission from Jun and Steinberg. Copyright 2011 by the American Physical Society.

Boundary layer in elastic turbulence

There is already some good data on the existence of a boundary layer in elastic turbulence (Jun and Steinberg 2011). Figure 5.4 shows data from particle imaging velocimetry measurements on polyacrylamide chains (long polyelectrolytes) that flow in small capillaries and experience elastic turbulence. Clearly there are regions near to the boundaries in which the average velocity experiences large perturbations.

References

Davidson P A 2015 *Turbulence: an Introduction for Scientists and Engineers* (Oxford: Oxford University Press)

Frisch U 1995 *Turbulence: The Legacy of A.N.Kolmogorov* (Cambridge: Cambridge University Press)

Groisman A and Steinberg V 2000 Elastic turbulence in a polymer solution flow *Nature* **405** 53

Guyon E, Hulin J P, Petit L and Mitescu C D 2015 *Physical Hydrodynamics* (Oxford: Oxford University Press)

Jun Y and Steinberg V 2011 Elastic turbulence in a curvilinear channel flow *Phys. Rev.* E **84** 56325

Malm A V 2015 Optical coherence tomography velocimetry and x-ray scattering rheology of complex fluids *Physics and Astronomy* (Manchester: University of Manchester)

Malm A V and Waigh T A 2017 Elastic turbulence in entangled semi-dilute DNA solutions measured with optical coherence tomography *Sci. Rep.* **7** 1186

Saddough S G and Veeravalli S V 1994 Local isotropy in turbulent boundary layers at high Reynolds number *J. Fluid Mech.* **268** 333–72

IOP Publishing

Chapter 6

How mucus works—the 21 mysteries in man

6.1 Introduction

Snot, slime, boggies and boogers; there is a huge variety of slang words used to describe mucus in the English language and this list is far from complete. The wide variety of names underlines mucus's important role in everyone's lives. An individual is in big trouble if their mucus stops functioning. They will be dead if there is a widespread breakdown, and will have a seriously compromised quality of life if more limited faults occur, e.g. the malfunction of respiratory mucins plays an important role in both cystic fibrosis (which has a genetic origin) and chronic obstructive pulmonary disorder (some similar symptoms to cystic fibrosis, but it is predominantly due to smoking). Gastrointestinal mucins also can malfunction in cystic fibrosis, which also contributes to morbidity. Nevertheless, just because mucus appears to be common-place in us humans, it does not mean its properties are well understood. As a rule of thumb, materials that appear mundane at first sight can often play an important role in people's lives, but surprisingly their exact functions are frequently overlooked. Any contribution to an improved understanding of mucus (even only minor improvements) could have a major impact on everyone's wellbeing and clearly research in this area should be encouraged.

Even mucin's structure, the major type of biomolecule that provides the physical properties of mucus, is still badly defined. It consists of a protein backbone that has carbohydrate side-chains attached to it with a comb architecture, but the exact arrangement of the side-chains is open for debate. Furthermore, the connection of the molecular architecture of mucin to its functional properties lacks a quantitative level of understanding. The two major classifications of mucin molecules are inventively called *gelling* and *non-gelling mucins*. Moreover, the name non-gelling mucin is used interchangeably, and slightly confusingly, in the literature with *membrane bound mucins* or *epithelial mucins*, since these molecules are always found chemically bonded to surfaces.

doi:10.1088/978-0-7503-1374-2ch6 © IOP Publishing Ltd 2017

In broad terms, the molecules of standard *gelling mucins*, such as those in the saliva on the tip of the reader's tongue, are known to consist of four components: a small globule, a big globule, a highly glycosylated comb region that connects the two globules together (the carbohydrate side-chains referred to previously are attached to this section) and a number of breaker regions (function unknown). Different mucin genes tend to give rise to slightly different molecular structures that are congruent to this general pattern. For gelling mucins, the small globules and big globules can stick to globules of neighbouring molecules that are similar to themselves, which creates a gelled material with elastic properties. To make matters even more complicated, both charged and uncharged versions can exist for the carbohydrate side-chains of mucins, which are chemically very heterogeneous, and this leads to added structural variability. At least the protein backbones of a particular type of mucin are all identical, which helps matters slightly, since they are all genetically determined and during their synthesis they are read directly from a person's DNA chains, whereas the carbohydrate side-chains can be affected by the person's diet.

Mucins are highly charged molecules (they are *polyelectrolytes*, chapter 17) and the charge interactions are long ranged, which means that a tricky many-body interaction determines their morphology (the many-body problem with charge interactions). For example, a charged sulphate group on one carbohydrate side-chain can interact with hundreds of other similar groups on the same molecule and also those on neighbouring molecules. This provides a big headache for prospective modellers. Currently no quantitative model exists that relates the molecular structure of mucin to the mechanical properties of its gels. Predictive mechanical models are in principle possible using modern theories of polyelectrolyte dynamics, but to date only modest advances have been made in their application to mucin.

In addition to the gelling mucins, there are also *membrane bound* (non-gelling) *mucins*, whose function is even less well understood than the gelling varieties. One important role for the membrane bound materials is that they form a protective brush that repels particles away from a cell's surface and they thus provide a physical defence mechanism on many internal surfaces inside the human body. For example, the gelling mucus on the lungs of humans is held off the cells' surfaces by an additional membrane bound layer of mucin molecules. The membrane bound mucins can help improve the motility of foreign particles inhaled into the lungs and the dust is thus easier to remove, since the membrane bound mucins obstruct mechanisms for these particles' adhesion. To counteract this repulsive interaction due to membrane bound mucins, some bacteria have evolved a wide range of molecular strategies that include lectins (proteins that can bind onto carbohydrates) to improve their chances of infection. The lectins glue the bacteria onto the mucins near to the environmental niche that they want to inhabit. Conversely, other mucin bound glues (such as IGA antibodies secreted into the intestines) form part of the human immune system, that permanently bind bacteria in a particular location to destroy their viability. Mucosal immunology is thus a large well-established field in biology and medicine, but a quantitative understanding of the phenomena involved is still in its infancy.

Before a good understanding of the function of mucins can be developed, it is necessary to have some good structural models.

L. *Question:* Deduce the molecular structure of all of the 21 human mucins with an Angstrom level of resolution.

There are a number of barriers that are presented to this challenge of structural elucidation. All the mucin molecules are non-crystalline, which obstructs the use of standard crystallography techniques, since the diffraction patterns become blurred. New experimental methods are needed, such as cryoelectron microscopy, super-resolution fluorescence microscopy or coherent x-ray imaging (chapter 8).

In the lungs, cilia (from the Latin word for eyelash) coat the airways, which are themselves covered with gelling mucin, and the cilia create Mexican waves of activity that transport dust particles upwards towards the mouth; a dust escalator that is vital for efficient breathing. Membrane bound mucins are also attached to the lung cell walls, as discussed previously, to prevent the adhesion of gelling mucus, dust and microbes to the cells' surfaces, which would obstruct the motion of the cilia. For the dust escalator to work well it is thought the gelling mucus layer needs a finite amount of elasticity, which provides something for the cilia to push against. The gelling mucus also needs to be fairly sticky to remove the dust particles and other pathogens (bacteria, viruses, fungi etc) that are blown into the airways. However, if the mucus becomes too elastic the escalator is compromised, cilia become damaged and patients constantly need to cough to remove lung pollutants; a relatively disruptive and inefficient mechanism in the lung's defences. This occurs in patients with cystic fibrosis (CF) and those with chronic obstructive pulmonary disorder (COPD). Related issues of mucus transport also happen in a wide range of other lung diseases, such as asthma and bronchitis.

Healthy lung mucus is important for all mammals and amusingly a molecular biologist in Manchester was recently funded by the Betting Agency to correlate the performance of horses with the state of their respiratory mucus. The Betting Agency is the regulatory body that controls horse racing in the UK. It is possible that a sputum sample could be used to determine the health of a horse and provide a better indicator of its capability of winning a race, which is why the research was funded. Research into lung mucus also has clear applications in veterinary medicine and the welfare of the animals.

Mucins play lots of other critical roles in the human body in addition to those in the lungs. In the stomach they form a protective layer that stops this organ from digesting itself. Evidently there is therefore a strong connection between the performance of the mucus layer on the walls of the stomach and the development of ulcers. Indeed, many stomach ulcers are now known to be caused by a bacterium, *Helico bacter* (Nobel Prize for medicine in 2005), that has evolved a corkscrew shape that is thought to facilitate its motion through the gelled layers of mucin found on the walls of the stomach. In most people, and the majority of the world's human population are thought to carry the infection, the bacteria cause no harm to the host, but in a small number of people they can cause stomach ulcers. Now that the causative agent for the disease is known, the treatment is fairly straightforward; the patient is treated with antibiotics that are targeted at the specific bacteria involved and in many cases they are cured of the ulcers. *Helico bacter* are now so well genetically characterised that the migration of modern man out of Africa around 100 000 years ago has been geographically mapped through the comparison of bacterial DNA from different people's stomachs. The story uncovered broadly follows that based on more standard evidence from archaeology and the genetics of human chromosomal DNA and mitochondrial DNA.

In the mouth, mucins are the most important molecules that form structures in saliva. Saliva helps to protect the teeth from decay due to bacterial growth. *Xerogenic* (dry mouth) conditions in people with throat cancer or amphetamine abusers reduce the production of saliva, and thus tend to be associated with tooth decay. People also find it difficult to swallow their food without saliva, which provides an unpleasant choking sensation, since saliva eases the passage of food down the throat. Purified (Orthana) mucin from the stomachs of pigs is commercially available as a pharmaceutical treatment for xerogenic conditions, although the development of other alternatives would clearly be useful.

Other more subtle issues with saliva are important for foodies and thus the food industry, such as *astringency*. Astringency is related to the puckering sensation felt in the mouth when some wines and tea are drunk. A major source of astringency is due to the aggregation of mucin in saliva. The biological physics group in Manchester was involved in some experiments on tea, in which the astringency was found to be connected to the action of specific polyphenol molecules (green tea contained the highest concentrations). Polyphenols also have beneficial antioxidant effects in humans, since they contribute to anti-cancer and to anti-ageing processes, so the aggregation of polyphenols by mucin should be avoided, or the polyphenols will not be absorbed into the blood stream to provide these possible benefits. Many polyphenols also have an antibiotic activity and those in tea are known to disrupt the formation of bacterial biofilms (chapter 10).

There are numerous gynaecological applications for mucin. These seem slightly comical in abstract terms, but can change the lives of childless couples. For example, sperm cells need to swim through a layer of mucus in order to achieve fertilisation of egg cells and subsequently these fertilised cells are moved down the Fallopian tube using a raft of cilia covered with yet another layer of mucus. The Fallopian cilia have a similar Mexican wave of activity to that used by the dust escalator created by cilia in the lungs. Picture a football stadium where the spectators propel a huge inflatable ball (the fertilised egg) around the stadium with a Mexican wave of coordinated activity. Quantitative descriptions of the motility of either sperm or fertilised eggs are still far from convincing, although they may be able to provide new insights into infertility in the future.

Mucins are also secreted in the eye to form a protective coating. This stops the eyelid shearing off a sensitive layer of the cornea each time someone blinks their eye. In the case of people who suffer from the condition of dry eye, the eyeball will become red and swell up. In extreme conditions this can lead to the necessity for a corneal transplant. Frictional effects on the surface of the eye are also very important for contact lens wearers and eye drops are available that partially replace the action of the mucins. Nevertheless, without a fundamental understanding of mucins' properties, researchers are not certain of exactly what they hope to replace, and biocompatibility is an issue if the behaviour of replacement materials markedly diverges from the naturally occurring fluids, i.e. the mucus created by the conjunctiva of the eye.

The genetics of mucin has steadily become better understood over the last 10 years, which gives some hope for future developments in related biomaterials problems at the molecular level. Genetic studies can allow the type of protein molecules formed by each mucin gene to be exactly defined, and their degree of expression in different parts of the body can also be determined. There are now known to be 21 mucin genes expressed in humans, but this number may increase slightly as new genes are discovered. The pig genome has also recently been published in the scientific literature, which helps push mucin research forward, since the pig is often used as a standard reference system, as it is hard to extract large amounts of mucin from living humans. Human saliva can be milked from the mouths of volunteers, but it varies by a surprising amount between individuals and considerable effort is required to purify large quantities of mucin via this method. Furthermore, human saliva only contains one or two of the possible 21 types of mucin that exists within people. Ideally the properties of all the mucins created by the different genes need to be known. Lots of pig mucin that is purified to a pharmaceutical grade is produced in Denmark, which is also famous for its bacon industry. This commercially available porcine mucin, as described previously, is used as a replacement material for people suffering from throat cancer and as a result this tends to be a model mucin system used in research laboratories.

The question of mucin's structure has already been posed, so the next challenge is to relate the structure of mucin to its function.

L. *Question*: Create a quantitative explanation of the functional role of the amino acid sequence of all the human mucins, or their equivalents in a pig.

More specifically, how are each of the 21 human mucin genes optimised for their specific duties? For example, why are there big variations in the length of intestinal mucins, whereas those in the stomach are much more similar in length? Why is a mixture of four different mucins used to coat the lungs of humans? Why are the globules in some mucins bigger than others? There are hundreds of unanswered questions.

At this stage the weird and wonderful uses of mucin in other non-human organisms will not be considered in detail, otherwise we will be overwhelmed by their vast range of uses. However, mucins do occur on the skin of dolphins and sharks to reduce turbulence and the build-up of microbial biofilms, in methods to capture prey by hagfish, the motility of slugs, in home building by swiftlets (found in birds nest soup) and the crystallisation of nacre in the shells of molluscs. There is also the related mucilage family of molecules used in pitcher plants to trap flies. Hagfish have a particularly gruesome application for mucin, because they catch their prey with a bag of mucin that is reinforced with keratin filaments. The hagfish thus create a slimy net based on mucin that is deadly for their prey.

A first step towards this goal of understanding human mucins would be to create a library of mucins' physical properties. The entire range of human mucins (due to the MUC genes 1 to 21) could thus be compared in a table in terms of their charge, viscosity, diffusion coefficients etc. This would need coherence between different experiments performed in different laboratories around the world to allow for the direct comparison of the results, e.g. similar temperatures, buffers, preparation procedures etc must be chosen. A very useful technique to study the physics of mucin has been the development of microrheology instruments that can explore the flow properties of microliters of specimen (chapter 13). Measurements of viscoelasticity with such instruments are prime candidates for their inclusion in the mucin library.

Gelation of a fluid has a dramatic effect on its physical properties. Think of creating a wobbly jelly in a kitchen for a children's party. The collagen is heated into a fluid state with the addition of lots of water and then cooled in a jelly mould. At room-temperature the fluid jelly will become solid-like at long time scales, although there is still a large amount of bound liquid water that is mobile at short time scales. Gels are therefore weird, slightly schizophrenic materials that are neither fully solid nor fully liquid. In polymeric solutions, such as mucin, gelation needs the introduction of cross-linkers (either intrinsic or extrinsic to the molecules) to create a network structure. How the network structures are formed in mucin is still not completely understood.

L. *Question*: Make an accurate quantitative model for the action of cross-linkers in mucin solutions that cause them to gel.

Cross-linkers are the names of the molecules that connect neighbouring polymers in a gel to create network structures. These cross-linkers are thus critical for the process of gelation and the resultant network architecture.

For example, with many of the so called *gelling mucins* (there is a hint in the name), when the pH of the solution is dropped it causes the molecules to spontaneously cross-link with one another to form a gelled network. Clearly this is very useful in the extremely acidic environment of the stomach to protect it from auto digestion (the mucin forms a protective layer on the stomach wall that stops it disintegrating under the attack of the digestive juices), but it also may occur for mucins localised in other organs. Thus, pH switchable gelation may be important in cystic fibrosis. The malfunctioning of ion channels in cystic fibrosis can drop the pH on the surfaces of cells in the lungs (and possibly the intestines), which causes gelation of the mucin layer and has medical implications due to its inhibited motility, e.g. the resultant mucus can no longer be easily mobilised by cilia in the lungs and as a result will damage them. Thick mucin gels are thus ideal biomaterials to protect the stomach, but could cause problems in many other environments, such as the surface of the lung.

Actin proteins provide vital scaffolds that form part of the backbone of human cells. With actin solutions there are now known to be over 50 specialized molecules that can create cross-links between the giant self-assembled actin filaments to create intracellular scaffolds. The cross-linker molecules can hold the actin chains perpendicular to each other, at a well-defined intermediate angle or hold them parallel in a bundle. Evidently the control of the cytoskeletal network architecture that is formed by the actin filaments is very important in cells. However, the analogous range of cross-linkers for mucins has not yet been catalogued in detail. To a first approximation the elasticity of gel networks is linearly related to the density of cross-links. Large numbers of cross-links hence lead to highly elastic networks. Therefore, if the process of mucin cross-linking can be better understood, rapid progress can in principle be made in terms of understanding the resultant elasticity and then viscoelasticity of the solutions. This could lead to a better understanding of pH switchable gelation of respiratory and intestinal mucins. Also, there is a wide range of naturally occurring proteins in humans that interact with mucins and many can act as cross-linkers (dubbed the *mucin interactome*). Furthermore, anti-nutrient molecules, such as lectins found in nuts, legumes and the coats of bacteria, are also thought to act as cross-linkers of mucin molecules. Bacteria thus tend to immobilise themselves in their preferred environmental niches in mucus, which can facilitate their proliferation and is a key first step in many human infections. Conversely, the immobilisation of bacteria in mucin gels by antibodies and lectins is also an important component of the human immune system that is used to fight infections. Trefoil lectins that are naturally secreted by the gut's immune system are known to cross-link mucins. The virulence of bacterial infections is often related to their degree of motility and their virulence is significantly impeded if they are trapped in a densely cross-linked mucin gel. The biochemical glues the bacteria create thus need to be carefully regulated and can be switched on or off depending on their needs.

Our own laboratory found that polyphenols in tea act as cross-linkers for mucins and a large range of other food molecules may also be able to demonstrate this behaviour. Anti-nutrient molecules form part of a vast arsenal of molecules that plants use to disrupt the health of animals that try to eat them. Among the molecules used are some of the most poisonous ones known to man, e.g. ricin is found in castor plants. This evolutionary pressure is so strong that the expectation is that we have hardly scratched the surface of the wide variety of naturally occurring molecules that will modify mucins.

As explained previously, the other main classification of mucins describes the epithelial (skin) tethered varieties. These charged combs are tethered to interfaces and tend to repel foreign molecules and pathogens from fragile cellular surfaces, but the exact forces involved are not well understood. Furthermore, it is expected that the interactions experienced by the surface-tethered brushes have an intricate interplay with the life cycle of pathogens and help to determine such processes as bacterial adhesion and the transmission of viruses between hosts. The majority of medically important viruses need to travel through mucosal surfaces for successful infections to occur and the field of mucosal immunology is well developed from a medical perspective, although the corresponding biophysics is mostly unexplored.

L. *Question*: Create a quantitative model for epithelial mucins (tethered brushes) attached to the surface of a human cell that is in reasonable agreement with experiments that measure the forces they experience, e.g. performed with atomic force microscopy.

Mucins are not the only comb polymers in the human body; proteoglycans, such as *aggrecan* and *versican* also play important roles. *Proteoglycans* are formed from proteins attached to a carbohydrate backbone, whereas the reverse is true for *glycoproteins* such as mucin. Aggrecan is found in cartilage where it acts as a composite with collagen. The giant aggrecan molecules have a very large viscosity due to their comb topology, which allows the composite material to dissipate substantial amounts of energy, and makes cartilage a very tough material. Replacement materials are needed for cartilage to treat its degradation in osteoarthritis. The hope is therefore to create replacement combs that emulate the natural behaviour of aggrecan (lubricin, a comb glycoprotein, is also now thought to play a role and it forms end-grafted brushes on the surface of cartilage), e.g. providing the same ultra-low friction coefficients that naturally occur in healthy synovial joints. Versican plays a similar role to aggrecan in the brain and helps to reduce concussive injuries to brain cells, e.g. in rugby, boxing or car accidents. Therefore, the entire range of these naturally occurring comb molecules needs to be better understood.

In conclusion, the current possibilities for large advances in man's understanding of mucins are very encouraging due to improvements in genetic techniques, physical models and high-resolution instrumentation. Fungus the Bogeyman is a fictional character created by Raymond Briggs who is covered in slime (with a daughter called Mucus) and is paid to scare normal humans. If we pay heed to Fungus the Bogeyman's words, '....just keep bogling away', mankind's understanding of mucin could improve in leaps and bounds.

6.2 Suggested reading

Introductory

Briggs R 2012 *Fungus the Bogeyman* (London: Puffin). An amusing book to entertain both children and mucin researchers alike.

Parham P 2014 *The Immune System* (New York: Garland). Mucus is of huge importance in medicine, since the majority of microbial diseases are transferred through a layer of the material during infections and a sizeable number of microorganisms spend their whole lives embedded in it.

Berg J M, Tymoczko J L, Gatto G J and Stryer L 2015 *Biochemistry* (New York: Freeman). Classic introductory textbook to biochemistry.

Advanced

McGuckin M A and Thornton D J 2012 *Mucins: Methods and Protocols* (New York: Humana Press). A major challenge in the field of mucin is to prepare well characterised samples and this book describes some of the modern biochemical protocols that are required. Many of the gelling mucin samples that have been studied in the literature do not actually gel due to problems with the sample preparation, which reduces their importance, since they are of more limited biological relevance.

Vaquez P A and Forest M G 2015 *Complex Fluids in Biological systems* (Berlin: Springer) pp. 53–110. A wide ranging mathematical discussion of mucosal surfaces.

Button B, Cai L H, Ehre C, Kesimer M, Hill D B, Sheehan J K, Boucher R C and Rubinstein M 2012 A periciliary brush promotes the lung health by separating the mucus layer from airway epithelia *Science* **337** 937–41. Includes a useful first attempt to describe the physics of epithelial tethered mucins. Much more detailed models are needed to make exact quantitative agreement with specific experimental systems.

Waigh T A and Papagiannopoulos A 2010 Biological and biomimetic comb polyelectrolytes *Polymers* **2** 57–70. An introduction to some of the polymer physics associated with the comb motif.

Shah V S *et al* 2016 Airway acidification initiates host defense abnormalities in cystic fibrosis mice *Science* **351** 503. This study implies acidification of the airways may play a significant role in human cystic fibrosis, since humans lack an ion channel that is present in mice that protects the mice from the disease.

Georgiades P, Pudney P D A, Rogers S S, Thornton D J and Waigh T A 2014 Tea derived gaolloylated polyphenols cross-link purified gastrointestinal mucins *PLoS ONE* **9** 1–11. The interaction of mucin with polyphenols from tea is explored to understand a possible cause of astringency.

Thim L, Madsen F and Poulsen S S 2002 Effect of trefoil factors on the viscoelastic properties of mucus gels *Eur. J. Clin. Invest.* **32** 519–27. Primary research literature that describes peptide crosslinkers in the intestinal immune system.

6.3 Technical details

6.3.1 Varieties of comb biopolymer

Figure 6.1 shows three examples of comb biopolymers. The range of possible molecular structures is very rich (Waigh and Papagiannopoulos 2010). Specifically, figure 6.1(a) shows the type of gelling mucins found in the stomach, (b) aggrecan (a proteoglycan) found in cartilage and (c) a non-gelling mucin found tethered to the surface of a cell. These diagrams are intentionally vague, since much of the exact molecular structure is incompletely understood.

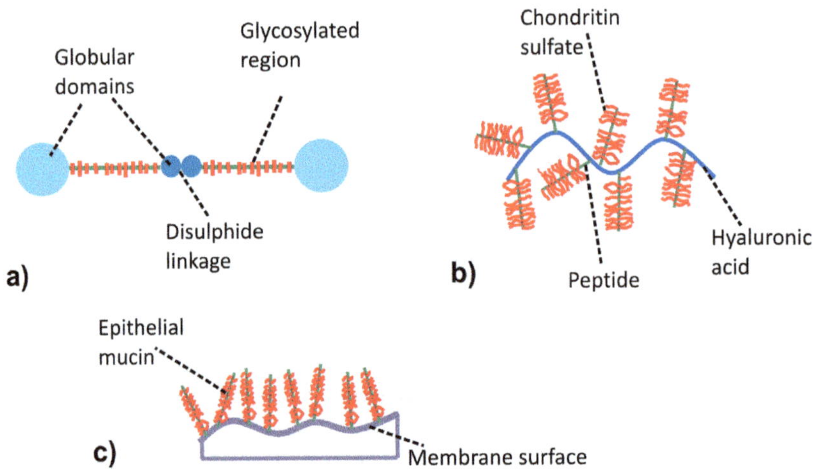

Figure 6.1. Three examples of comb biopolymers, (a) gelling mucins (MUC5AC), (b) aggrecan and (c) epithelial mucins (Waigh and Papagiannopoulos 2010).

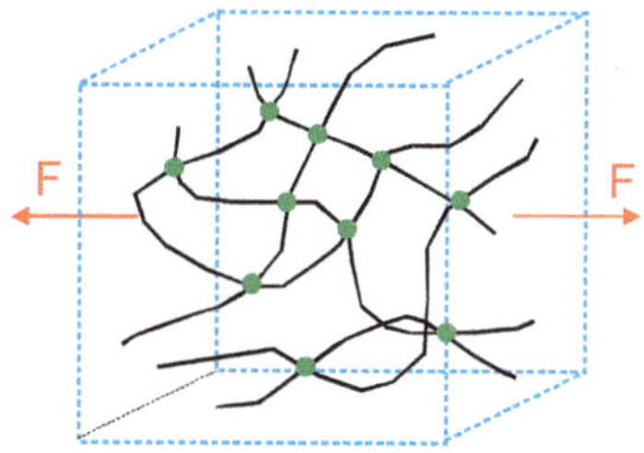

Figure 6.2. An elastic cube of cross-linked polymers under tension (force, *F*). To a first approximation the elastic shear modulus of the cube is given by equation (6.1), i.e. the shear modulus is proportional to the cross-link density and the coordination number of the junctions in the network. The cross-linkers are indicated by green circles.

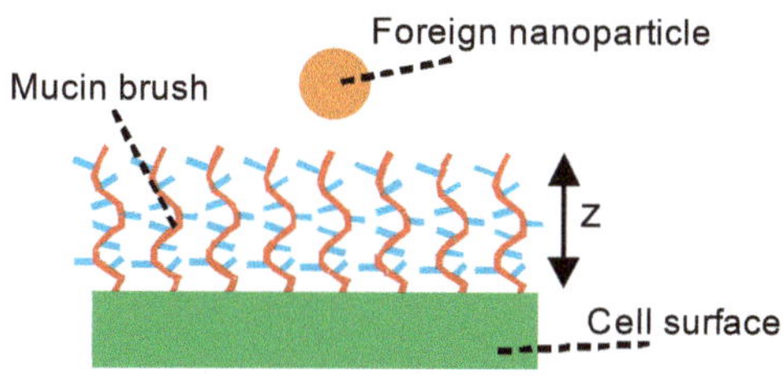

Figure 6.3. Surface tethered polymer brushed (a model for epithelial mucin) can exert a repulsive force on a nanoparticle.

6.3.2 Gelling mucins

The elasticities of gels are often well described by an extremely simple calculation, which could be extended to a description of mucin gels (figure 16.2). The elastic shear modulus (G) is linearly proportional to the density of cross-links, with a constant prefactor that depends on the average number of chains connected to each cross-link,

$$G = C\mu kT \tag{6.1}$$

where kT is the thermal energy (k is Boltzmann's constant and T is the absolute temperature), C is a factor related to the average coordination number of the cross-link and μ is the average cross-link density. Thus, characterisation of the cross-linking density is a critical parameter to understand the mechanical properties of mucin gels and it can be modulated by molecules in the mucin interactome, e.g. polyphenols in tea.

6.3.3 Epithelial mucins

Polymer scaling theories can be used to calculate the concentration profile of polymer brushes near to a surface (figure 6.3), e.g. a model for epithelial mucins. Button *et al* derive a simple expression for this concentration profile ($\phi(z)$) as

$$\varphi(z) = \left(\frac{b}{\xi}\right)^{4/3}\left[\log\left(\frac{z_0}{z - z_0}\right)\right]^{-4/3} \tag{6.2}$$

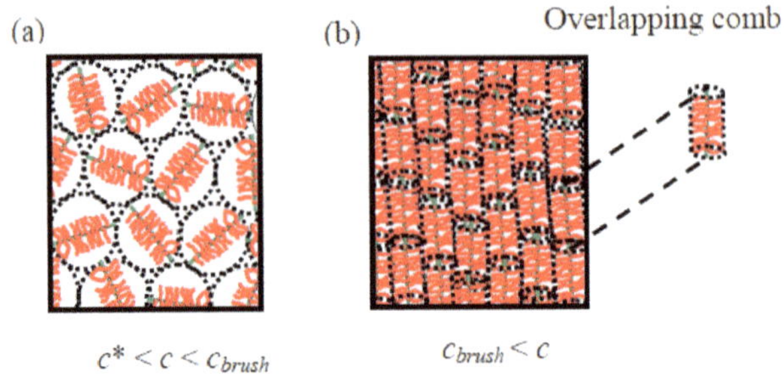

Figure 6.4. (a) c^* is the conventional semi-dilute overlap concentration of the comb backbone. (b) A brush concentration (c_{brush}) occurs with single comb polyelectrolytes, such as mucin, at high concentrations (Waigh and Papagiannopoulos 2010).

where z_0 is the height of the brush, b is the monomer step size in the mucin and ξ is the semi-dilute correlation length (Button *et al* 2012). Similar models can be extended to calculate the force on a nanoparticle brought into contact with the brush.

6.3.4 Viscosity

The viscosity of a solution of polyelectrolyte combs, such as gelling mucins, is a sensitive function of their architectures. Our group derived some expressions to describe the viscoelasticity of simple flexible polyelectrolytes combs (a simple extension of the Dobrynin model for linear polyelectrolytes) (Dobrynin *et al* 1995, Papagiannopoulos *et al* 2005). For example, the specific viscosity (η_s) for unentangled semi-dilute solutions of polyelectrolyte combs is

$$\eta_{\mathrm{sp}} \equiv \frac{\eta - \eta_s}{\eta_s} \approx \frac{K^2}{N_{\mathrm{tot}}}\left(\frac{B}{b}\right)^{-3/2}\left(1 + \frac{n^2 s}{K(B/b)^{3/2}}c^{1/2}\right)c^{1/2} \tag{6.3}$$

where η is the comb viscosity, η_s is the viscosity of water, c is the mucin concentration, and K, N_{tot}, B, n and s are related to the geometry of the comb. The idea of the brush overlap concentration is incorporated into such calculations (figure 6.4), which occurs in addition to the standard semi-dilute and the entanglement concentrations that occur with flexible polymers (Rubinstein and Colby 2003).

6.3.5 Aggrecan

Aggrecan has a more complicated bottle brush of bottle brushes morphology, which can self-assemble from its components (figure 6.5) (Papagiannopoulos *et al* 2008). In terms of the dynamics of aggrecan, three overlap concentrations can be defined: the conventional semi-dilute overlap concentration and two brush overlap concentrations, i.e. an additional overlap concentration is defined compared with mucin. Chain overlap plays an important role in theories for polymer dynamics, because it helps to determine the effects of hydrodynamic screening.

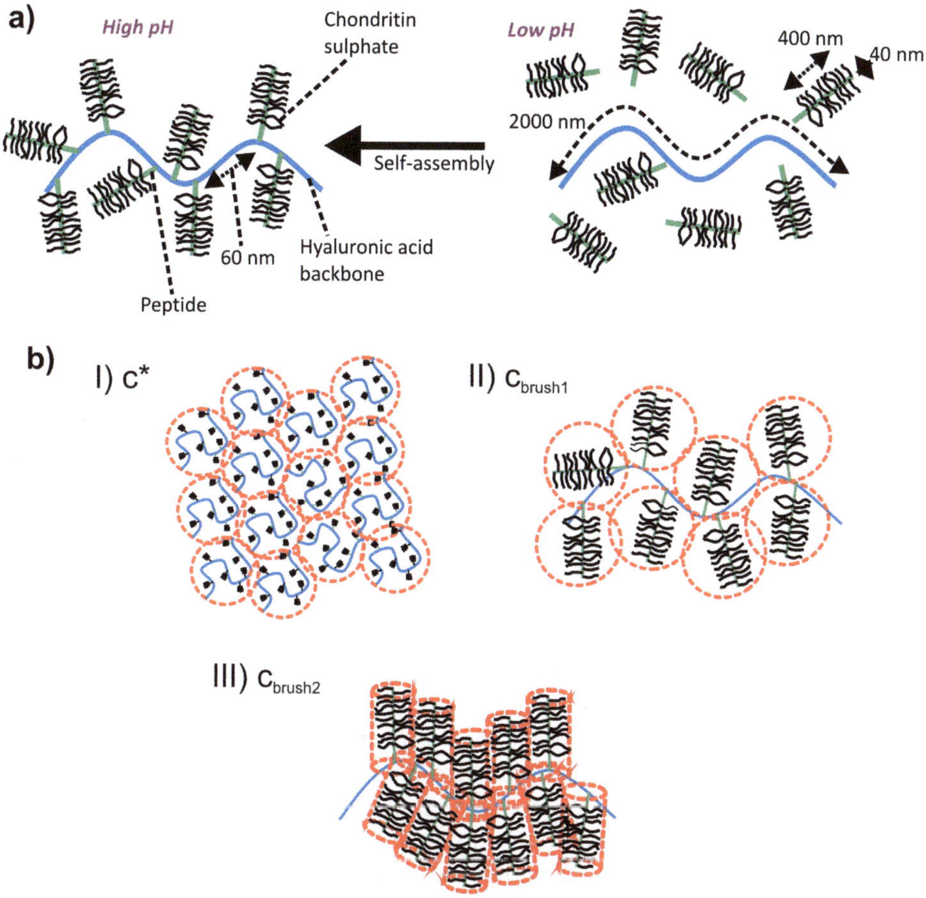

Figure 6.5. (a) In aggrecan the brush monomers can self-assemble onto a hyaluronic acid backbone to create bottle-brush of bottle brushes architecture. (b) Due to the hierarchical structure of aggrecan, three overlap concentrations can be defined, (i) c^* is the semi-dilute overlap concentration, (ii) c_{brush1} is the first brush overlap concentration and (iii) c_{brush2} is the second brush overlap concentration (Papagiannopoulos *et al* 2008).

References

Button B, Cai L H, Ehre C, Kesimer M, Hill D B, Sheehan J K, Boucher R C and Rubinstein M 2012 A periciliary brush promotes the lung health by separating the mucus layer from airway epithelia *Science* **337** 937–41

Dobrynin A V, Colby R H and Rubinstein M 1995 Scaling theory of polyelectrolyte solutions *Macromolecules* **28** 1859–71

Papagiannopoulos A, Fernyhough C M and Waigh T A 2005 The microrheology of polystyrene sulphonate combs in aqueous solution *J. Chem. Phys.* **123** 214904

Papagiannopoulos A, Waigh T A and Hardingham T E 2008 The viscoelasticity of self-assembled proteoglycan combs *Faraday Discuss.* **139** 337–57

Rubinstein M and Colby R H 2003 *Polymer Physics* (Oxford: Oxford University Press).

Waigh T A and Papagiannopoulos A 2010 Biological and biomimetic comb polyelectrolytes *Polymers* **2** 57–70

IOP Publishing

Chapter 7

Synthetic biology—reengineering bugs and molecules

7.1 Introduction

Famously, Ernest Rutherford is quoted as saying, 'There is only physics and stamp collecting'. This is often used by physicists to poo poo other areas of scientific enquiry, and they frequently direct their criticism at colleagues from chemistry departments. A slight irony is that Rutherford won his Nobel Prize in chemistry, not physics. In this the physicists are perhaps motivated by some unpleasant formative experience of chemistry at school, when they failed to remember a reaction scheme correctly in front of the class, did badly in an exam or made some nasty smells when they mucked up an attempt to synthesise some new molecules. Nevertheless, their exact motivation for this chemistry phobia is incidental, and complete ignorance of chemistry has a number of risks associated with it. A language without any syntax is just a lot of blah blah blah, and chemistry is a major component of the syntax of the real world. Detailed knowledge of quantum field theory (the semantics of ultra-small-scale phenomena) will not, for example, help someone decide whether a malfunctioning protein is implicated in a certain disease, but a good working knowledge of organic chemistry might do. Chemistry thus forms some of the most important syntax that underlies the language of biological physics.

A modern paradigm in chemistry is *synthetic biology*, and it is such a good test bed for biological physics that many physicists have recently jumped aboard the band wagon. In a nutshell, synthetic biology describes the creation of both new life forms and new biological molecules that have not been seen previously in Nature. Clearly ethical considerations are important here, but to date only fairly modest tweaks to existing organisms have been possible and the evolutionary arms race experienced by all the other pre-existing life forms has caused all of them to become very stream-lined, which literally means the majority of these new life forms would not last 10 minutes in the world outside their laboratories. That is, if they were able

doi:10.1088/978-0-7503-1374-2ch7 © IOP Publishing Ltd 2017

to escape. The new synthetic organisms would quickly become meals for leaner and meaner life forms that already inhabit the modern world. To avoid even this possibility, one fail safe measure is to insure the new organisms are sterile. A controversial example is the sterile seeds that have been introduced into genetically modified wheats to stop the escape of the plants into their surrounding environments and to prevent inter-breeding with other wild species of grass. Farmers are unhappy, because the grains produced cannot be used in subsequent years, which ties them to additional purchases from agrochemical companies.

Whilst sterilisation is frequently possible with higher organisms, it is very hard to achieve with bacteria, since they can swap DNA extremely successfully with their neighbours. Care is thus taken in all respectable laboratories to avoid that the escape of bacteria happens in the first place, and none have yet been documented. Furthermore, it must be stressed that nothing radical has yet been done with synthetic life forms, in part due to the major limitations presented by current technologies, so no man-eating Triffids or superbugs have been created. Triffids appeared in a fiction book written by John Wyndam. They were carnivorous plants designed to solve the world's energy crisis through the synthesis of petroleum using photosynthesis, and the plants created had the unfortunate defect of having a penchant for human flesh. Thankfully Triffids remain a subject of pure science fiction, but there are risks associated with tinkering with microbes that are already extremely capable of causing disease in humans, e.g. the genetic modification of the smallpox virus. Such research on microorganisms is thus carefully controlled by responsible governments.

Synthetic biology promises some neat solutions to classic biophysics questions, such as the components necessary to create a minimal genome for a living cell, e.g. the shortest possible length of DNA that is required. Another example is a method to study the inner workings of biological machines, such as the flagellar motor on a bacterium, by tweaking the associated proteins in the genome. The length of the flagellum (whip-like appendages which spin around to propel the bacterium) could be varied to observe its effect on thrust or membrane proteins could be adapted to modify the hydrogen ion concentration that drives the motor. A further example would be to create a mutant mouse with ion channels in its brain that could be switched on or off using a pulse of light (this field is called *optogenetics*). The impact of the ion channel's activity on the brain of the mouse could be observed and its effect on diseased states could be deduced.

There is conceivably lots of money to be made from synthetic biology solutions to real world problems that confront society, either with the new life forms constructed or the new molecules that they in turn can create. For example, the insulin needed to treat diabetics or human growth factor supplements for people with developmental conditions can be mass produced using re-engineered bacteria. Both applications are now famous successful case studies for genetic engineering and they have improved the quality of life of many individuals, e.g. a human form of insulin is commercially available that is produced by genetically modified bacteria, whereas previously diabetics needed insulin from pigs that was not an exact match for the human version and thus had an increased health risk (the insulin injections could cause additional immune responses in the patients).

Another fascinating possibility is the use of synthetic biology techniques in the *human genome write* project, which should not be confused with the *human genome read* project. The human genome read project was completed in 2003 and it created a complete map of all three billion base pairs in the human genome. The 2016 write project in contrast aims to synthesise an entire human genome from scratch using the fundamental base pair components and to then express them in cultured human cells. It must be stressed that no suggestion has been made of actually creating a fully synthetic human being and it would still be far from a practical possibility, even with a new synthesised genome. The flexibility in the design of these newly written genomes would then allow the role of regulatory RNA (up to 98% of the genome) to be explored and thus understood in much more detail, which would provide vital insights for disease (chapter 11). *De novo* (completely new) cancer cells could also be created from first principles, to form a better understanding of their metabolism and thus the molecular mechanisms important in the disease. Furthermore, the development of quantitative models for the RNA would be given a big boost as a systematic range of modified genomes could be constructed, to pinpoint the key features that control the dynamics of the processes involved.

The synthetic biology of completely new biological molecules can also be considered. Fortunately, many of the challenging parts of the chemical synthesis of many varieties of biological macromolecule have been removed, as designer peptides and nucleic acids can be mail ordered from chemical companies at reasonably low costs and with reasonably high purities. This is a major advantage for researchers who do not have a natural flare for chemical synthesis. The onus is now on the designer of the molecules to come up with a clever idea for the sequence of a macromolecule on a computer, relatively unconstrained by the technical requirements of the processes of organic synthesis that are necessary to make the actual molecules in a laboratory.

In terms of the regulation of the environment on mankind's hospitable blue planet, microorganisms currently do most of the hard work. They recycle people's effluents, create vital molecules for agriculture and help control the atmosphere, to mention only a few of their all-pervading roles in planet management. There is thus a clear precedent for the performance of useful jobs by microorganisms that could be harnessed using methods from synthetic biology, although large-scale planet management is not an obvious first target for this research due to the enormous scales involved. One more practical aim is the creation of synthetic biomaterials, and bacteria have consequently been used to synthesise plastics that are biodegradable, e.g. polylactic acid pellets can be created by bacteria, that can be subsequently processed into compostable yoghurt pots.

An early example of synthetic biology, this time from a physics perspective, was the creation of the *repressilator* in *E. coli* cells (a standard type of bacterium that lives inside human intestines). A new genetic circuit was introduced into *E. coli* cells that consisted of three genes that each repressed one of their two neighbours. Once inserted in the cells, the concentrations of the three proteins transcribed by the genes constantly oscillated inside the bacterium; a continuous game of 'rock, paper, scissors'. One of the genes was tagged with a fluorescent protein (originally found in jellyfish), so the 'rock, paper, scissors' game caused the bacteria to constantly flash on and off when viewed under a fluorescence microscope when the fluorescent protein was expressed (the movies can be Googled on the Web). More impressively, the dynamics of the new gene circuit were in good agreement with a mathematical model (that consisted of two coupled non-linear differential equations) to describe the oscillations.

An additional specific example of modern synthetic biology is given by the synthesis of peptides (primitive short chain proteins). The demarcation between a *peptide* and a *protein* is fairly arbitrary. Both contain the same 20 types of amino acid glued together with the same type of peptide bond. When less than 50 amino acids are connected the molecule is called a peptide and chains with more than 50 are called proteins. The main advantage of peptides is that they are easier to synthesise due to their small size and simple folding patterns, and they are thus cheaper. Dependent on their exact sequences, peptides can be very biologically active (e.g. they can specifically kill bacteria or act as neurotransmitters) or they can be used as relatively inert structural materials. After many years of development, synthetic peptides materials are now a profitable industry and medical applications of peptide pharmaceuticals already create revenues on the order of billions of pounds.

Non-pharmaceutical uses of peptides are slightly rarer, although there appear to be many possible markets for structural peptide materials. An example of a commercial peptide biomaterial is the self-assembled peptides that are available for the media used to culture cells, i.e. porous scaffolds to encourage cell growth. This is a relatively niche market, but larger non-pharmaceutical markets may become open to peptide materials as methods to process them become better controlled and production costs become cheaper, e.g. textiles made from spider silks created by genetically modified bacteria. Many of the current stumbling blocks for the profitability of such research are in the chemical engineering aspects of materials processing, i.e. how to scale up the processing conditions of biomaterials to make them economically viable. Spider farms that are arranged as high-throughput factories are impractical for silk production, since spiders have lots of antisocial habits at high population densities, e.g. cannibalism is rife among spiders. Spider silks that have been created by genetically modified bacteria have been woven into rugs, ties, Adidas trainers (sneakers; a very expensive limited edition) and used as violin strings, although these bacterial strategies have not yet made the transition to mass production, in part due to challenges in materials processing.

There are also lots of elegant examples of molecular synthetic biology with DNA origami, where the chains fold themselves spontaneously into the correct shapes using the process of self-assembly, e.g. many millions of identical DNA chains have each been self-assembled into individual smiley faces that are all one hundred nanometres in diameter. The utility of smiley faces may not seem immediately obvious, but they act as a proof of principle and many more useful alternatives are possible using the same methods that have been developed. A correct DNA sequence can be designed on a software package and then mail ordered to a researcher's laboratory, which will automatically fold itself into a functional nanostructure when dissolved in water, e.g. to produce an artificial ion channel to clean dirty water. DNA can also be used as a medium for information storage and it has been used to store recorded music. Nanomachines can be made with DNA origami that consume energy and produce useful work, e.g. stepper motors that walk along surfaces. These provide fascinating systems to test our understanding of non-equilibrium thermodynamics at the single molecule level. The processes of DNA folding are better understood than that of protein folding, which facilitates these origami designs, but the chemistry is slightly more limited with DNA, e.g. only four basic flavours of nucleic acid base occur with fairly similar physical properties, compared with the twenty varieties of amino acid possible with proteins, which are physically much more varied.

After reviewing the three principle types of synthetic biology, based on bacteria, peptides and DNA, we are now in a position to pose a very general question.

L. *Question*: Create a simple synthetic biology design to make yourself lots of money and solve a real world problem.

In the biological physics group in Manchester synthetic biology designs entail the use of bacteria or peptides, rather than DNA origami, due to the areas of expertise currently possessed by staff members. There have already been some successes in our laboratory. A specific Mancunian example of synthetic biology is cationic peptides for the disruption of bacterial membranes that avoid killing neighbouring human cells, while destroying the troublesome bugs. These peptides could be used as a novel broadband antibiotic for wound dressings. Indeed, natural inspiration for this solution is seen in frogs and insects that lack such a sophisticated immune system as humans, and primarily rely on secreted peptides to control infections. Other new Mancunian peptides can act as efficient blood clotting agents to immediately stem the flow of blood in an operation, e.g. a peptide coating sprayed onto a wound during an operation might be possible. However, care is needed to stringently test inventions with medical applications and both these types of peptides, although they show some promise, are still far from achieving the FDA or the EMFA approval needed for them to be routinely used by clinicians. Much more extensive investment of time and money would be required, and the failure rate for these types of inventions is often very high. From a physics perspective we are still lacking detailed insights into exactly how the Mancunian peptides function, e.g. how the cationic peptides modify the curvature of bacterial membranes to cause the cells to explode or how the blood clotting peptides can be used to create mechanically strong sealants (surgeons often use super-glue in operating theatres and a more biocompatible substitute is required).

Every year there is a synthetic biology competition to create genetically modified bacteria held for undergraduate students called iGEM (international Genetically Engineered Machines) that originally took place in Boston (USA), but is now a truly international event with hundreds of different nations that participate. The organisation that runs the competition also provides a convenient registry of more than 20 000 genetic parts, which enables students to create new organisms over a relatively short period of time, e.g. a new bug can be designed and tested during a two month summer project. Past winning projects include the invention of photographic films based on light sensitive bacteria, bacteria that create perfume, bacteria that perform water purification..., the number of possibilities are currently more limited by the imagination of the designers, rather than any real technical restrictions.

The ability to genetically modify organisms to mass produce new molecules means synthetic biology has become a modern paradigm in biomaterials research for the production of green ecologically friendly bulk materials, which will have a huge range of potential applications. The field of regenerative medicine is rapidly evolving, driven by the needs of our aging population, which requires the creation of replacement body parts for the old, injured or sick. In the future, it is possible that the field of synthetic biology will fulfil many of the needs of regenerative medicine for new biomaterials. Synthetic biology also provides a powerful set of tools to explore the underlying physical principles that need to be understood to optimise its roles in many new applications.

7.2 Suggested reading

Introductory

Church G M and Regis E 2012 *Regenesis* (New York: Perseus). An enthusiastic popular description of recent advances in synthetic biology.

Advanced

Lilijeruhm J, Gulberg E and Forster A C 2014 *Synthetic Biology a Lab Manual* (Singapore: World Scientific). Required reading for entrants into the iGEM competition to create new synthetic life forms.

Seeman N C 2016 *Structural DNA Nanotechnology* (Cambridge: Cambridge University Press). An up-to-date textbook on DNA nanotechnology from a leader in the field.

2017 iGEM. (igem.org: MIT). A website that describes the rules of the international genetically engineering machine competition (iGEM). It is particularly useful for prospective participants who want to begin a new design using the genetic modification of *E. coli*.

Boeke J D, Church G, Hessel A and Kelley N J 2016 The genome write project *Science* anf6850. This article makes the case for the creation of a completely synthetic human genome.

Bai J K, Chen C X, Wang J X, Zhang Y, Cox H, Zhang J Z, Wang Y M, Penny J, Waigh T A and Lu J R 2016 Enzymatic regulation of self-assembling peptide nanostructures and hydrogelation with highly selective antibacterial activities *ACS Applied Material and Interfaces* **8** 15093 Primary research literature that considers the synthetic biology of new peptides to treat bacterial infections.

Chen C *et al* 2016 Hydrogelation of the short self-assembling peptide I3QGK regulated by transglutaminase and use for rapid hemostasis *ACS Applied Material and Interfaces* **8** 17833–41. Primary research literature on the use of synthetic biology to create new blooding clotting peptides.

Elowitz M B and Leibler S 2000 A synthetic oscillatory network of transcriptional regulators *Nature* **403** 335–8. Primary research literature that describes the invention of the repressilator circuit and its expression in live *E. coli* cells.

Cohen A E and Venkatachalam V 2014 Bringing bioelectricity to light *Ann. Rev. Biophys.* **43** 211–32. Describes recent advances in optogenetics to control the response of ion channels.

7.3 Technical details

7.3.1 Genetic engineering

The field of genetic engineering is primarily based on the *central dogma of biology* as proposed by the biophysicist Francis Crick. It can be simply defined by the interaction of three types of molecule with a one-dimensional flow of information,

$$\text{DNA} \rightarrow \text{RNA} \rightarrow \text{protein}$$

Within this simple model (which is still thought to hold true to a good approximation with a few exceptions related to the field of epigenetics) all the information in a cell is found within its DNA. Chains of RNA are transcribed from the DNA and subsequently the RNA chains are translated into proteins. Proteins do most of the construction jobs at the molecular level inside cells, e.g. they form nanofactories (enzymes). Artificial modifications of DNA cause different proteins to be expressed and thus lead to different behaviours of the cell. There are some caveats; since the cell needs to be expressed inside a living cell to fully exploit the information contained and cannot be thought of as completely independent of the cell type. A physics PhD student in our laboratory, with no prior biological training, took around six months to arrive at the stage where he could correctly modify a single protein in *E. coli* (we were interested in the pathways used to synthesise capsules in disease causing *E. coli*, chapter 10).

7.3.2 How to engineer a bacterium

The standard technique currently used to engineer bacteria is called *gene cloning* (Dale and Parks 2010). The devil is in the detail with this type of methodology, so careful reference needs to be made to the primary literature to make the technology work in a real laboratory, since bacteria are fairly fussy about the conditions that are used. The basic steps involve cutting and joining DNA using restriction endonucleases to insert a new section of DNA into a plasmid. The DNA is often inserted

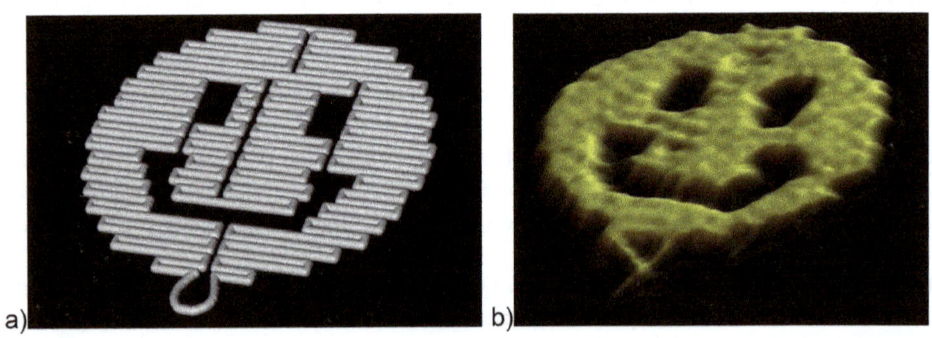

Figure 7.1. (a) The design of DNA origami for the self-assembly of a smiley face. (b) A TEM image of an actual self-assembled smiley face made from DNA (Rothemund 2006). The diameter of the face is ~100 nm. Reprinted by permission from Macmillan Publishers Ltd.

with a selectable marker (e.g. antibiotic resistance), so only the genetically transformed bacteria are amplified (i.e. during cloning).

7.3.3 How to engineer DNA

DNA can be self-assembled into designed nano-constructs using base pairing (Seeman 2016) (figure 7.1). DNA comes in four different flavours A, G, C and T. Famously in a double helix G pairs with C (forming three hydrogen bonds) and A pairs with T (forming two hydrogen bonds). Nevertheless, there are many other molecular structures that can be stabilised using the correct sequences of oligonucleotides and the good news is the rules that determine them are fairly well understood.

7.3.4 How to engineer peptides

Peptides are constructed from amino acids which are available in 20 different standard flavours and it is relatively straightforward to create short sequences using solid state synthesis methods (Vrachen and Weiss 2013). For example, with peptide surfactants the number of alanine units has a big effect on the resultant self-assembled structures (figure 7.2). As the number of alanine units increase the structure varies between (a) stacked bilayers, (b) nanofibers and (c) nanorods.

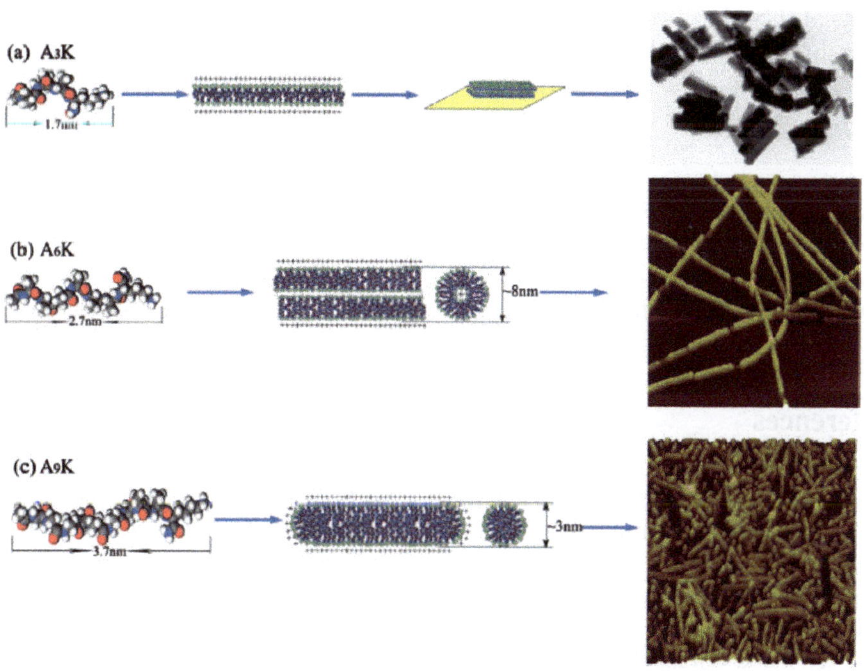

Figure 7.2. Cationic peptide surfactants (short mixtures of two types of amino acid) can self-assemble into nanostructures. A is alanine and K is lysine. These peptides have applications as broadband antimicrobials. Reprinted with permission from Xu *et al* 2009. Copyright 2009 American Chemical Society.

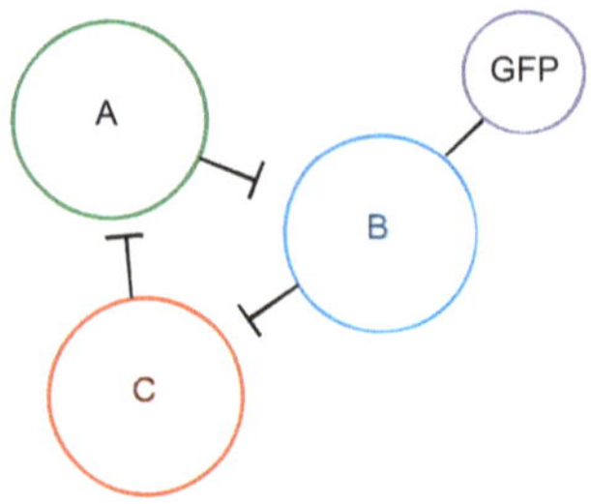

Figure 7.3. The repressilator is a classic example of synthetic biology (Garcia-Ojalvo *et al* 2004). Three genes, A, B and C, are inserted into the genome of a bacterium and repress one another in a circle, analogous to a continuous 'Rock, paper, scissors' game. If one of the genes products is tagged with GFP, the bacteria expressing the repressilator will blink on and off.

7.3.5 The repressilator

A schematic diagram for the gene circuit for the repressilator is shown in figure 7.3 (Garcia-Ojalvo *et al* 2004). The mathematical description of the repressilator is based on two coupled non-linear equations.

$$\dot{x}_1 = \kappa_1 f(x_2) - \gamma_1 x_1 \tag{7.1}$$

$$\dot{x}_2 = \kappa_2 f(x_1) - \gamma_2 x_2 \tag{7.2}$$

where x_1 is the concentration of the gene product form by A and x_2 combines the effects of the other two genes (it simplifies the models by lumping them together). γ_1 and γ_2 are constant rates of degradation. κ_1 and κ_2 are positive constants and $f(x)$ is an expression for the enzyme reaction rates,

$$f(x) = \frac{\theta^n}{\theta^n + x^n} \tag{7.3}$$

which is the Hill equation for the enzyme activity, where θ is a constant and n is a cooperativity parameter (Waigh 2014). Since analytic solutions are not possible for equations (7.1) and (7.2) when n is reasonably large, only numerical solutions can be explored, guided by analytic calculations for their stability and graphical descriptions of the behaviour of their phase space, e.g. plotting x_1 versus x_2.

References

Dale J W and Parks S F 2010 *Molecular Genetics of Bacteria* (Chichester: Wiley)

Garcia-Ojalvo J, Elowitz M B and Strogatz S H 2004 Modelling a synthetic multicellular clock: repressilators coupled by quorum sensing *PNAS* **101** 10955–60

Rothemund P W K 2006 Folding DNA to create nanoscale shapes and patterns *Nature* **440** 297–301

Seeman N C 2016 *Structural DNA nanotechnology* (Cambridge: Cambridge University Press).

Vrachen D V and Weiss G 2013 *Introduction to Bioorganic Chemistry and Chemical Biology* (New York: Garland Science)

Waigh T A 2014 *The Physics of Living Processes: a Mesoscopic Approach* (Chichester: Wiley)

Xu H, Wang J, Han S Y, Wang J Q, Yu D Y, Zhang H Y, Xia D H, Zhao X B, Waigh T A and Lu J R 2009 Hydrophobic-region-induced transitions in self-assembled peptide nanostructures *Langmuir* **25** 4115–23

IOP Publishing

Some Critical Questions in Biological Physics
A guided tour around the bugbears
Thomas Andrew Waigh

Chapter 8

Missing instruments—grease monkeys required

8.1 Introduction

To provide an entertaining perspective on scientific research, the four main types of scientist will be defined in terms of the least popular aspects of their professions: *cleaners*, *grease monkeys*, *dancers* and *watchers*. The *cleaners*, *grease monkeys*, *dancers* and *watchers* paradigm also provides an alternative entertaining scheme for understanding how people fit into the bigger picture in research collaborations that involve multiple disciplines. People who actually make something are either cleaners or grease monkeys.

To create or manipulate molecules a scientist finds themselves in the realms of chemistry (synthetic chemistry, biochemistry, forensic chemistry etc). For the chemist to have a high yield of their molecule they need to be very clean and very careful. Thus, historically a chemist's job involved cleaning lots of glassware, particularly when things often went wrong, and the chemist was left with a black mess stuck to the bottom of their beaker, due to the unwanted product of a side reaction. The beaker needed to be cleaned, the errors identified and then the process of synthesis restarted. Similar issues with cleanliness occupy significant amounts of time for bacteriologists (the so called aseptic technique needed to ensure that rogue bugs do not confuse their results), nanotechnologists (boiler suited individuals inside their clean rooms that worry about dust in their transistors), medics (hand washing, mask wearing surgeons with sterilized scalpels), veterinarians, viticulturists… the list is huge. Good *cleaners* are thus needed by a huge range of research establishments. It is stressed that these are tongue in cheek descriptions of negative aspects of research professions, since professional scientists are predominantly very earnest, clever and hard working. Many of them also have very flexible skill sets.

Biophysics without any quantitative models is extremely unsatisfying. In the absence of any numerical descriptions the level of understanding can seem fairly superficial. However, theoretical physics divorced from reality (without any reasonable physical assumptions) becomes more like pure mathematics; it sometimes

doi:10.1088/978-0-7503-1374-2ch8

© IOP Publishing Ltd 2017

appears more of an art form than a science. Theorists can resemble *dancers*, where on-lookers can only comment on the high quality of their technique, their pirouettes and their choreography, but the equations derived will have a negligible impact in the real world, and are of little relevance to biological physics.

An amusing competition for scientists has been held for a number of years in which entrants create a dance based on the subject matter of their PhD thesis, whose results are often reported in the prestigious journal *Nature*. The author is himself a poor dancer with two left feet, but he did once have a post-doc who ran a Lindy Hop class in the evenings with his wife and he was also a very competent researcher in the world of optoelectronics, so clearly these two skill sets can coexist in a single individual.

For *watchers*, the realms of observational science are being described. Historically much of biology was observational, and painstaking descriptions of the behaviour of plants and animals were made; these are the traditional fields of botany and zoology. Think of an entomologist in a jungle with a butterfly net or Jacques Cousteau scuba diving in his amazing undersea world. Observational research also plays an important role in astronomy, where it is almost impossible to perturb the systems investigated to any significant degree. In some respects people should probably be thankful that black holes or supernovae cannot yet be created in the laboratory, and measurements with telescopes are always uninvasive in astronomy. For example, imagine the number of health and safety forms that would need to be completed before a new supernova could be created. Nevertheless, Richard Feynmann famously said, 'What I cannot create I do not fully understand', and this is always a major problem with observational science, which makes it slightly unsatisfactory. It is hard to develop a thorough understanding of a phenomenon without perturbing it in some manner, to decipher the individual contributions of all the competing interactions, and people are restricted to a form of scientific voyeurism. In this Heisenberg's uncertainty principle is being ignored, but perturbations to measured systems are innate to all possible measurements. It is written into the fundamental laws of nature, although the perturbations to astrophysics phenomena by telescopic observations are expected to be unmeasurably small.

The construction of missing scientific instruments is mostly a job for *grease monkeys*. *Grease monkey* is used as a generic description of a wide range of engineering disciplines. Some people use *grease monkeys* specifically for mechanical engineers, *sparkies* for electrical engineers (a badge of honour used to be the number of times they had electrocuted themselves) and *planters/plumbers* for chemical engineers (plants are required to mass produce chemicals and they involve lots of pipe work). Grease monkeys are the type of people who can construct a Mach–Zender interferometer in their sleep, know their power amplifiers from their lock-in amplifiers and can demonstrate zen-like degrees of patience when trouble shooting intermittent faults in vast complicated electrical circuits. Provided someone has this type of skill set (or is willing to learn), they can start to consider gaps in the market place for measurement apparatus in biophysical research.

The good news is that there is plenty of important work in instrument design still left to be done and a steady stream of new high specification components continue to become available (particularly in photonics) that will be able to completely disrupt the influence of current technologies, i.e. technologies are becoming possible that will completely revolutionize areas of research. For example, orders of magnitude improvements can be implemented with new improved light sources, such as LEDs and lasers, e.g. using frequency comb lasers with ultrahigh sensitivity. These new components can provide higher fluxes of light over a wider range of wavelengths and can be constructed at cheaper costs than previously possible. There are numerous other improvements in photonic instrumentation, such as active optics, infrared, terahertz and fibre based components that are also now available. In many fields the new technologies will provide orders of magnitude improvements in resolution and make invisible signals visible. People are therefore encouraged to dust off their copies of Horowitz and Hill, load up their grease guns and prepare for some action.

In terms of patentability, and thus the prospects for the creation of money from biophysics, the statistics are skewed very strongly in the order: cleaners have better prospects than grease monkeys, who in turn have better prospects than dancers and watchers, i.e. there are a reasonable number of rich chemists and engineers, and relatively few affluent theoreticians, zoologists and botanists. Computer programmers (*coders*) tend to sit outside of this classification scheme. If the software can help the cleaners or grease monkeys to complete their tasks, it may well make a little money in biophysics. Thus, in conclusion, not only are there lots of jobs for grease monkeys, there are also reasonable profits to be made constructing new instruments in biological physics. For the more entrepreneurial minded readers, reference should also be made to chapters 7 and 16 for some other money-making opportunities, this time more closely aligned with the field of chemistry.

So, what is on the shopping list for new instrumentation in biological physics? There are lots of unfilled niches in the market place.

In general terms, people tend to want four things from instruments in biological physics: high molecular specificity, high spatial resolution, high temporal resolution and low damage to the specimens. Clearly there is lots of auxiliary equipment that does not snuggly fit into this simple classification scheme, such as centrifuges and electrophoresis gels (they only have high molecular specificity), but here we are considering the principal instrumentational workhorses used in molecular biophysics. Often these competing requirements for an instrument are antagonistic to one another, e.g. the resolution can be increased through an increase of the radiation dosage, but samples are inevitably fried in the process. Ideally, the measurements need to leave the organism, cell or molecule completely unharmed or they risk irrecoverably perturbing the living processes that are the subject of the investigation. Such issues have become more and more relevant in the 21st century, as people want to move beyond the simple structural elucidation of molecules, to properly understand their action in living processes. Invasivity also provides a clear barrier for many physical experiments in medicine, e.g. problems with ionising radiation are a constant cause for concern in hospitals with x-ray imaging experiments and those with positron emission tomography (PET) scanners (γ-ray photons).

There still exist many missing experimental windows into the behaviour of biological materials over specific time scales and length scales, e.g. the provision of high resolution imaging in thick opaque sections of living tissue is currently very limited. Biophysics experiments also need extreme specificity, since there are a heck of a lot of different biological molecules that are created in living organisms and people want to be certain of the ones that are observed, to help construct accurate molecular models of their activity. This is a major advantage of some standard biophysical techniques, such as fluorescence microscopy, nuclear magnetic resonance (NMR) spectroscopy and Raman spectroscopy, i.e. they can provide molecular specific information. As a result these three methods are widely used. NMR also forms part of the magnetic resonance imaging (MRI) technique that is discussed later in this chapter.

New photonics components that can be mass produced will help drive the development of disruptive technologies within the field of biophysical instrumentation. The data created using photonic devices built with new improved components can suddenly become orders of magnitude better resolved and new windows into different time and length scales become open, which in turn can allow totally new biological phenomena to be observed. Nanophotonics instruments also offer some huge advances when efficiently integrated with biological systems, not only in terms of resolution, but also in terms of the minute quantities of specimen that are required and their ability to function well in high throughput applications, e.g. when used in microrheology apparatus to measure the viscoelasticity of minute specimens (chapter 13).

Often it is hard to join the skill sets of individuals who can build a piece of scientific equipment with those who can actually make use of the apparatus to solve a current biological problem. Optical breadboards resemble Meccano sets where the interchangeable components can be easily reconfigured. Optical physicists can therefore appear like ice sculpture artists whose designs are constantly disintegrating as the equipment is cannibalized for the next bigger and better design. Real ice sculpture artists use chain saws to create sculptures from large blocks of ice, but the art form is delicate and transient, and the masterpieces rapidly melt in the Sun. This transience can be frustrating for a biophysicist who needs a robust well defined piece of instrumentation to be available, to help them to solve a particular problem.

8.1.1 Raman spectroscopy

Some promising experimental methods will now be described in more detail. First *Raman spectroscopy* is considered, named after its Indian discoverer who, incredibly in the 1920s, made the first experiments with just sunlight focused by some lenses, a liquid sample in a glass container (60 different common liquids all showed similar effects), some filters and his own eyes. Standard Raman scattering is an extremely weak process and in modern laboratories requires extremely sensitive detectors combined with powerful lasers for its detection. In what is now known as Raman scattering, a small amount of light that is scattered by a sample experiences an energy shift, and the energy shifts are characteristic of the specimen. The measurement of these energy changes means that the technique is a type of spectroscopy. The specific definition of Raman spectroscopy is a little more subtle, in that it involves scattering induced by changes in the polarizability of a specimen and the energy changes are characteristic of the molecules' internal vibrations and rotations, but these details will not concern us further here. Suffice to say Raman spectroscopy is an extremely useful method to fingerprint the type of molecule found in a biological specimen by scattering light off it.

An ideal biophysical experiment would be based around spectroscopic imaging in a microscope, in which the molecular composition of every component that is imaged could be identified. A first target for such a microscope in biology would be to explore a single living cell.

L. *Question*: Create a Raman microscope based on a high power pulsed laser (a non-linear optical effect can then occur due to the high fluxes of photons) that provides readily decipherable data on the molecules inside live cells at millisecond time scales, with high spectral resolution and relatively low photon dosages.

The emphasis on decipherable data with this microscope is required, since it is preferable to avoid that the spectra is mangled in a non-linear manner, or detailed analysis of the molecular structure will become a big challenge. Once complete, the new Raman microscope could allow the diagnosis of cancerous cells on a cell-by-cell basis with extremely high fidelity. In a clinical setting, the cells could be provided by minute biopsies from tumours found in patients or the light scattered from them could be directly measured in the body through the insertion of an optical fibre. Many of the currently used Raman microscopy methods have a high possibility of false positives in cancer detection due to the weak Raman signals involved, since Raman scattering is a million times weaker than the standard scattering processes normally used to create images in optical microscopes. Consequently, Raman scattering is not a standard technique for cancer detection in hospitals. The emphasis on non-linear Raman microscopy is made, since these new techniques, that involve powerful pulsed lasers, can be used to amplify the weak signals involved to provide high resolution spectra from single cells. There are hundreds of other issues in cellular biology in addition to cancer that could be revolutionised with spectroscopic imaging techniques that have reasonably high resolution, such as the networks of information transfer inside human cells that are described in chapter 1.

8.1.2 Magnetic resonance imaging

Magnetic resonance imaging (MRI) is a fantastic technique to probe living organisms. It is nearly completely non-invasive and, as yet, there are no known health implications associated with high levels of exposure to the fields involved in magnetic resonance imaging apparatus (unlike x-ray radiation used in CT scans or gamma ray radiation in PET scans). Extremely high magnetic fields, higher than those used in MRI, can disturb mental processes in individuals, but the effects appear to be fairly short term. This technique of *transcranial magnetic stimulation* is thus now relatively well studied in medicine and has been trialled as a treatment for depression and Asperger's disease. The high magnetic fields and radio waves (that contain very low energy photons) necessary for MRI experiments, thus appear to do little long term damage to living organisms. MRI instruments can function across entire organisms (whole people) and can provide on the order of 10 micron resolution images across their whole bodies. It is clear why hospitals and laboratories around the world continue to invest large sums of money to achieve state-of-the-art MRI facilities (for example the University of Manchester just bought another machine for £5M), since they can play a crucial role in the diagnosis of disease.

A large component of the costs for MRI is the superconducting magnets, because the resolutions achieved on the images scale with the magnetic field strength, with a few technical ifs, buts and maybes. Super-conducting magnets need to be kept at very cold temperatures, which means the running costs to supply the coolant gases can also be very expensive. To build large continuous magnets with field strengths much above one Tesla, superconductors are always needed, which leads to the necessity for these very expensive devices. Improvements in low field MRI, which uses cheaper magnets that can only create low magnetic field strengths, would thus provide huge benefits for medical science. Machines would become readily portable and their price would dramatically reduce, so MRI machines could be used for routine investigations in doctor's surgeries or in ambulances rather than their restricted use in hospitals. Good progress has been made from both the perspective of cheap high field magnets and sensitive low field detection with MRI, so there is some hope that a practical solution can be achieved. Single molecule sensitivity in NMR has also recently been demonstrated with the use of defects in diamonds, although it is not yet possible in living cells.

L. *Question:* Create a high resolution magnetic resonance imaging device with much lower requirements on the strength of the magnets than is currently possible.

Image contrast is also an issue with MRI, since it is needed to allow specific *in vivo* processes to be followed. Functional MRI (sensitive to the binding of oxygen to haemoglobin) has been used to map mental activity inside live human brains and has become a key tool in brain research. However, the 10 micron resolution that currently can be achieved in brain research is far away from the 10 nanometre resolution (a thousand times smaller) required to observe single synapses, and images of synaptic networks are needed for a fully reductionist model of brain activity (chapter 3). Clearly much more work is required to provide a fully satisfactory solution for brain imaging.

8.1.3 Super-resolution fluorescence microscopy

Super-resolution fluorescence microscopy was the subject of the Nobel Prize for Chemistry in 2014 and the microscopes continue to become better and better. A key insight was that the resolution limit of standard optical microscopes (called the diffraction limit) can be broken by manipulating the quantum mechanics of fluorescent molecules with lasers. This has led to an order of magnitude improvement of the resolution from hundreds of nanometres to tens of nanometres. Thus, it has suddenly become possible to image many structures inside intact cells for the first time, without any invasive procedures for sample preparation. Furthermore, there appears to be lots of scope for additional improvements in spatial resolution and dynamic range, and the technology has not yet struck any hard, unyielding technical barriers that will obstruct its progress. A big improvement would be to increase the time resolution of the techniques, since they are currently very slow, e.g. five minutes are required to collect an image of a single cell at 20 nanometre resolution using the STORM technique. The acronym STORM stands for STochastic Optical Reconstruction Microscopy.

L. *Question*: Construct a super-resolution fluorescence microscope that has 20 nanometre spatial resolution and millisecond time resolution.

Preferably a microscope should be built that can meet these resolution criteria in space and time, without burning the sample to a crisp using a high power laser. This goal of fast intracellular imaging has nearly been achieved and will revolutionise cellular biology when it becomes routinely available to a large user community. The current state of the art for super-resolution fluorescence microscopy allows the creation of a single image with around 50 nanometres resolution over a time period of 40 milliseconds (the Eric Betzig group). Improvements in the quality of the fluorophores used to label samples will also have a big impact on the field, since image resolutions often scale with the photon yield of the fluorophores before they are photo-bleached by the lasers used to switch them on, which permanently switches them off.

8.1.4 Optical coherence tomography

Next there is a question in the fluid mechanics of biological materials that would profit from further consideration of the suitable instrumentation.

L. *Question*: Map the flow patterns in three dimensions for an opaque biological fluid with high temporal and high spatial resolution using the technique of optical coherence tomography.

Optical coherence tomography (OCT) is a relatively new imaging technique, and currently the main commercial application for it is in ophthalmology for imaging the retina at the back of the eye. The big advantage of OCT is its ability to explore opaque materials, e.g. retinal imaging is possible when the cornea at the front of the eye has been damaged to make it opaque. The majority of other optical techniques are confounded by a small degree of sample opacity. To date OCT apparatus have mostly been optimized for imaging, but it can also be used to measure sample velocities using Doppler shift measurements. Light frequencies change slightly when scattered from a target that moves, similar to the change in frequency of a locomotive's whistle as it passes a train station at speed. There are lots of applications for OCT in biological fluid mechanics, if it can be optimized for dynamics with high temporal resolution. The biological physics group in Manchester have a patent on the combination of OCT with a rheometer (the scientific equivalent of a food blender that is used to study viscoelasticity, chapter 13), that has many applications for the exploration of the flow patterns of viscoelastic materials. Many unusual medically important flow instabilities could be probed for the first time in opaque fluids using this technique, e.g. the dynamics of blood relevant to its motion in the heart, the flow of mucus important during a cough in the lungs or the motion of DNA that is used in high throughput microfluidic devices.

8.1.5 Cryo transmission electron microscopy

Cryo transmission electron microscopy (TEM) has made some big advances over the last few years. Its main drawback is that only dead samples can be investigated due to the strong interactions of electrons with biological specimens and sample preparation is as a result fairly invasive. Samples need to be chopped up into thin slivers, rapidly frozen (using liquid ethane at −188 °C) and then placed in an ultrahigh vacuum. Nevertheless, once specimens have been correctly prepared, it is possible to construct Angstrom resolution images of non-crystalline specimens in three dimensions, which gives TEM unparalleled strengths in many areas of structural biology, since this level of spatial resolution is unprecedented in imaging experiments. The number of cryo electron microscopy structures published in biology has consequently taken off exponentially over the last few years. A major practical challenge is to make the technique as easy as possible, so the method becomes available to the largest possible community of researchers.

8.1.6 Coherent x-ray imaging

Coherent x-ray imaging techniques have been mainly developed on synchrotrons (large expensive donut shaped particle accelerators for electrons) and free electron lasers (large expensive linear accelerators for electrons), but if the technique could make the transition to hospitals it would have a very large impact in medicine. For example, the development of phase contrast imaging with coherent x-rays in the future would enable a dentist to use x-rays to measure soft tissue, as well as bone; to check for gum disease as well as damage to the teeth. Currently only the teeth are visible in standard incoherent x-ray imaging methods that only provide absorption contrast. The increases in contrast possible on images created with coherent x-rays imply that dosages can be reduced, which for example would help for the screening of patients with breast cancer. High x-ray dosages can cause cancers in addition to those the imaging technique is being used to detect and therefore the reduced dosages are a big bonus. The main challenge is thus to create a budget source of coherent x-rays, since synchrotrons and free electron lasers are very bulky and cost upwards of £100M.

L. *Question*: Create a coherent x-ray source that is compact (table top), reasonably priced and has a bright well-collimated beam.

The next challenge is to use this coherent x-ray source in medicine to diagnose patients with lower dosages of x-rays than previously possible, e.g. to monitor bone fractures, investigate gum disease or for the screening of breast cancer. This transition to a doctor's surgery involves the standard barriers that limit translational research; the facility needs to be robust, stable and safe before anybody considers its use to image anything biological, let alone a medical application with patients in a hospital. Röntgen, the discoverer of x-rays, was particularly cavalier during his first experiments to explore the medical uses of x-rays, e.g. he used them to image his wife's hand without any idea of the extreme risks involved with ionising radiation.

The actual monetary cost has reduced significantly for some new prototype coherent x-ray sources. For example, the Technical University of Munich in Germany recently installed a €5M machine, which encroaches on the sums of money available at the hospitals of large European cities. Such budget coherent sources would be widely used in medicine, if their safety and utility could be comprehensively demonstrated.

8.1.7 Context

Once any of these possible apparatus have been built as functional prototypes, there is still a significant amount of work required to make them successful in the market place. Additional engineering is needed to develop a robust cost effective instrument from a working prototype. Conventionally there are lots of Heath-Robinson aspects to all functional science prototypes. Think of lots of crucial bits of Gaffer tape and blu tack that are integral to the prototype that need to be replaced before successful commercialisation is achieved. Furthermore, the scientific community needs to be properly educated on the capabilities of new methodologies. People require lots of effort to master new tools and this effort needs to be minimised, if the tools are to become really popular. The manufacturer has also to establish a good reputation for servicing the apparatus and dealing with customers. Inevitably software is required to run the equipment that should be as simple as possible, to provide non-experts with dependable measurements, even if they have a relatively shaky grasp of the physical principles that underlie them (*idiot proof* is the relevant engineering terminology). Clearly the creation and development of new instruments is not a task taken lightly, but big rewards are offered by successful designs (either monetarily or in terms of the improved quality of life facilitated by the new improved inventions).

In more general terms there is a real need for physicists to become engaged in up-to-date biological research problems. To be led by the biological aspects of a problem may feel awkward and unnatural to researchers originally trained in the physical sciences. However, crucial contributions have been made and will continue to be made by people willing to connect with the hot topics the biologists are genuinely interested in.

8.2 Suggested reading

Introductory

Horowitz P and Hill W 2015 *The Art of Electronics* (Cambridge: Cambridge University Press). A classic tour of modern electronics for the design of physical instruments.

Hobbs P C D 2009 *Building Electro-optical Systems: Making it all Work* (Chichester: Wiley). A useful introduction to practical aspects of optoelectronic designs.

Serdyuk I N, Zaccai N R and Zaccai J 2007 *Methods in Molecular Biophysics: Structure, Dynamics, Function* (Cambridge: Cambridge University Press). An expansive review of experimental techniques in molecular biophysics.

Belcasto S M and Schaffer K 2012 *The Best Writing on Mathematics 2012*; Pitici M (Princeton, NJ: Princeton University Press). This essay considers the overlap of dance as an art form with the science of mathematics.

Leake M 2016 *Biophysics Tools and Techniques* (Boca Raton, FL: CRC Press). A good introduction to the physical techniques actually used by molecular biologists and cellular biologists.

Advanced

Waigh T A 2014 *The Physics of Living Processes: A Mesoscopic Approach* (Chichester: Wiley). Chapter 19 gives a compact introduction to a range of important experimental techniques in biological physics.

Raman C V and Krishnan K S 1928 A new type of secondary radiation *Nature* **3048** 501–2. Primary research literature that describes the first experimental demonstration of Raman scattering that was performed using sunlight as the light source and the experimenters' own eyes as the detectors.

Waigh T A and Rau C 2012 X-ray and neutron imaging with colloids *Current Opinion in Colloid and Interface Science* **17** 13–22. Describes some new developments with coherent x-ray imaging with colloidal nanoparticles (both biological and synthetic).

Rust M J, Bates M and Zhang X 2006 Sub-diffraction limit imaging by stochastic optical reconstruction microscopy (STORM) *Nature Methods* **3** 793–5. Primary research literature that introduces a key super-resolution fluorescence microscopy technique.

Li D *et al* 2015 Extended-resolution structured illumination imaging of endocytic and cytoskeletal dynamics *Science* **349** 944. A current state-of-the-art super-resolution microscope is described from the group of the Nobel prize winner Eric Betzig.

8.3 Technical details

8.3.1 Raman spectroscopy

Raman spectroscopy involves the inelastic scattering of photons from a sample. The material perturbs the energy of the scattered beam and it can be explained in terms of virtual energy states (figure 8.1). This can be contrasted with fluorescence, which involves real excited energy states. In order for a molecular rotation or vibration to be Raman active a component of the molecular polarizability must change during the scattering event (Ferraro 2002).

8.3.2 Non-linear Raman spectroscopy

A key issue with standard Raman scattering is that it is an extremely weak process. Pulsed lasers increase the peak optical power experienced by a sample and can give rise to multi-photon Raman processes (figure 8.2). These in turn can amplify non-linear Raman effects, such as coherent anti-Stokes scattering and stimulated scattering, and lead to spectra with improved signal-to-noise ratios (Masters and

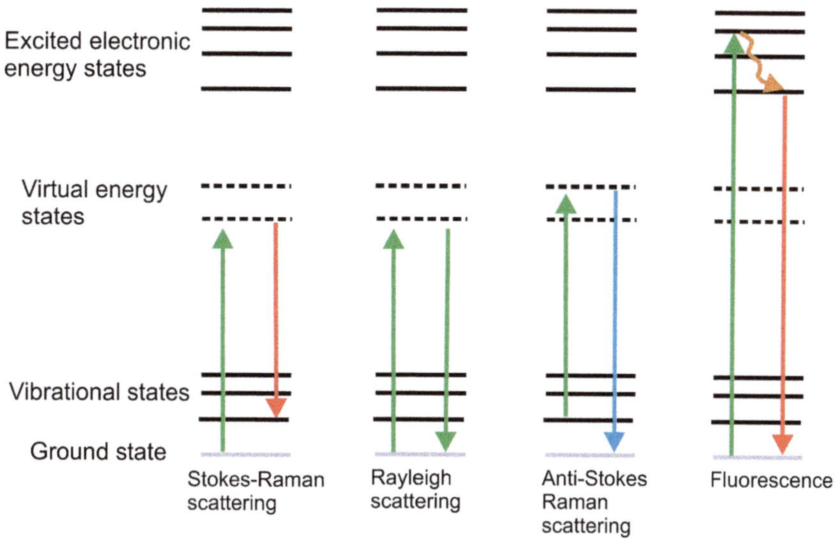

Figure 8.1. The energy transitions for molecules that experience two types of Raman scattering (Stokes and anti-Stokes) are compared with Rayleigh scattering and fluorescence emission.

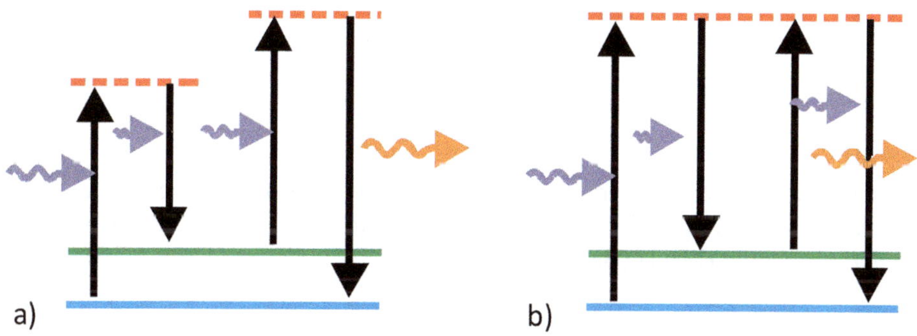

Figure 8.2. More sophisticated types of Raman scattering are possible using multi-photon processes that are facilitated by intense pulsed lasers and they are non-linear optical phenomena. (a) Coherent anti-Stokes Raman scattering and (b) stimulated Raman scattering.

So 2008). Thus, high resolution spectrally resolved images of live cells have been obtained.

8.3.3 Magnetic resonance imaging

Nuclear magnetic resonance (NMR) can be considered as a nuclear Zeeman effect. The application of a magnetic field lifts the degeneracy in the energy levels of the nuclei (figure 8.3). The separation of the energy levels (hf) is equal to that carried by radio frequency photons and this enables a form of absorption spectroscopy in which the photon frequency is varied to detect the nuclear resonances (Callaghan 1993). NMR can then be extended to provide an imaging capability by using a gradient in the magnetic field across the sample to detect the spatial position using

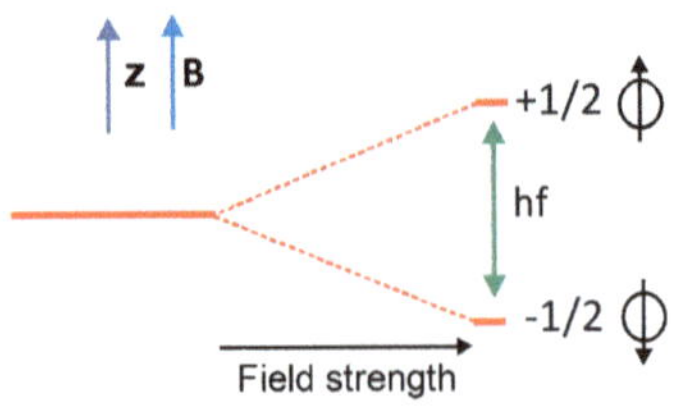

Figure 8.3. Nuclear magnetic resonance is based on a nuclear Zeeman effect due to the application of a magnetic field (B) in the direction z. The nuclear energy levels are shown with continuous lines. hf is the energy of a photon. ±½ are the spin on the nucleus with respect to z. Variation of the field strength causes the separation of the energy levels to vary, which can be used to measure the spatial position of the nuclei.

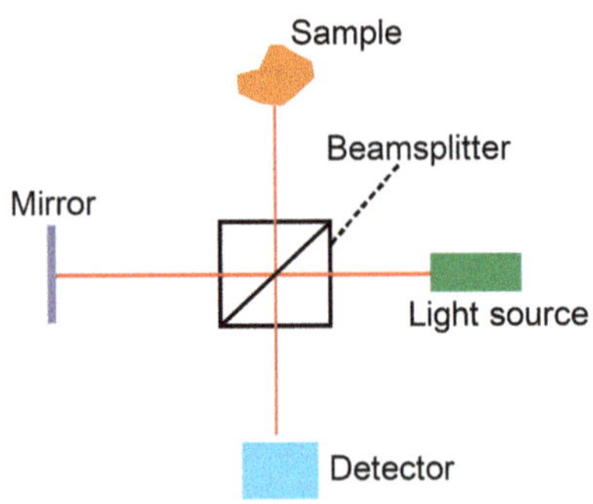

Figure 8.4. Schematic diagram of a Michelson interferometer used for optical coherence tomography (OCT) imaging.

variations in the energy splitting. Accurate application of the magnetic field gradients allows imaging in three dimensions and the technique is called *magnetic resonance imaging* (MRI) (Buxton 2009).

8.3.4 Optical coherence tomography

The simplest incarnation of optical coherence tomography (OCT) is built around a Michelson Interferometer (Drexler and Fujimoto 2015) (figure 8.4). The same type of device was used by Michelson and Morley in 1887 to show that the velocity of light (c) is a constant, which was subsequently used as the basis of Einstein's theory of special relativity. More recently, similar apparatus has been used in the detection of gravitational waves due to the collisions of black holes. Using infra-red light, such interferometers can allow imaging well inside opaque samples (~2 mm human tissue). If the sample moves during an experiment, the fringes on the detector due to interference of the two beams also move. The velocity of the sample can thus be calculated using an accurate analysis of the fringe motion (a Doppler effect). An application of Doppler OCT has been developed in our own laboratory for the motion of complex fluids in a rheometer. It allows analysis of the motions inside opaque fluids as they are sheared, e.g. *wall slip* (the sample moves as a plug, slipping on the walls) tomato ketchup samples or *shear banding* (stripes form with different velocities) in polyacrylamide solutions (figure 8.5).

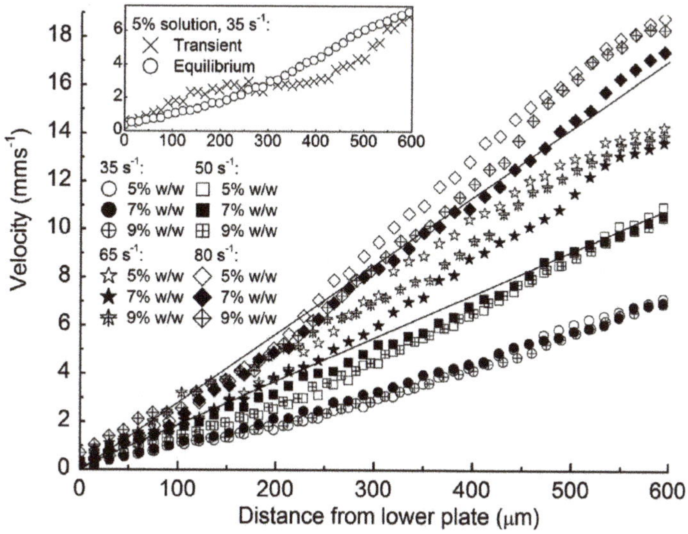

Figure 8.5. *Shear banding*, a non-linear flow instability, can be measured in a rheometer using Doppler optical coherence tomography (Jaradat *et al* 2012). For Newtonian fluids a constant velocity gradient is predicted. Shear banding describes a curvature in the velocity profile (alternatively described as strain localization).

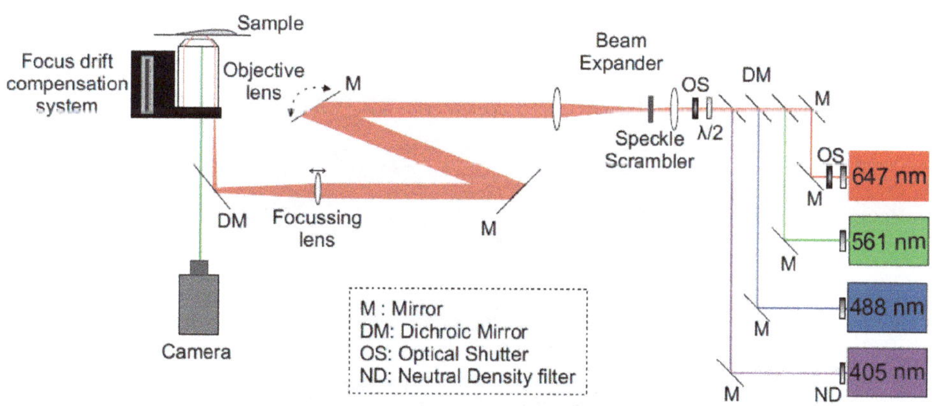

Figure 8.6. Stochastic optical reconstruction microscope (STORM) built in the University of Manchester laboratory (Georgiades *et al* 2016). Four lasers (647, 561, 488 and 405 nm) can be used to switch fluorophores on and off, so their positions can be accurately measured. A speckle scrambler is placed in the beam path to reduce coherent artefacts in the images, i.e. fringes.

8.3.5 Super-resolution fluorescence microscopy (STORM)

Super-resolution fluorescence microscopy is possible in a range of forms, but here the STORM technique will be discussed, a formidable acronym, that stands for STochastic Optical Reconstruction Microscopy. The apparatus has been constructed in our own laboratory (figure 8.6) and it is based on a design of Zhang's group at Harvard (Rust *et al* 2006).

The diffraction resolution limit (d) for a microscope can be calculated from a theory due to Abbe,

$$d = \frac{\lambda}{2n \sin \theta} \tag{8.1}$$

where $n \sin \theta$, the numerical aperture of the objective lens, is normally on the order of 1 for the lenses used on modern optical microscopes and λ is the wavelength of the radiation used (Mertz 2010). Therefore, the Abbe equation gives a limit of around 200 nm for visible light. A physical insight (due to Betzig) was that the resolution limit is due to the overlap of the emission of fluorophores used to label samples in both time and space. If the fluorophores can be switched on at different times, the resolution limit given by equation (8.1) can be broken. STORM experiments thus function with blinking fluorophores that are densely attached to the biological structures of interest. At one specific point in time only ~10% of the fluorophores are switched on. The centres of these fluorophores can be located to around ~10 nm resolution, and images of the sample are slowly built up in a dot-to-dot fashion. Pictures resemble the pointillist style of early 20th century impressionist painters, e.g. Georges Suerat. A super-resolved image of microtubules inside cells made with STORM is shown in figure 8.7 (the dots are imperceptible in the zoomed out view).

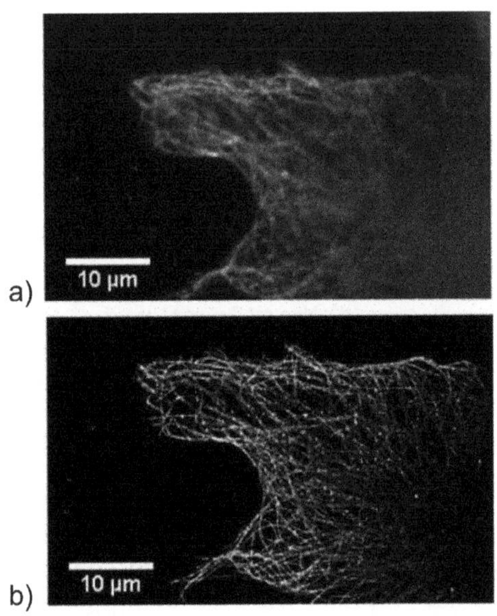

Figure 8.7. The microtubule network inside a human lung cell labelled with specific fluorophores (antibodies connected to dye molecules) can be image using fluorescence microscopy. a) A conventional fluorescence microscopy image with 250 nm resolution. b) A super-resolved STORM image with 30 nm resolution (Georgiades *et al* 2016).

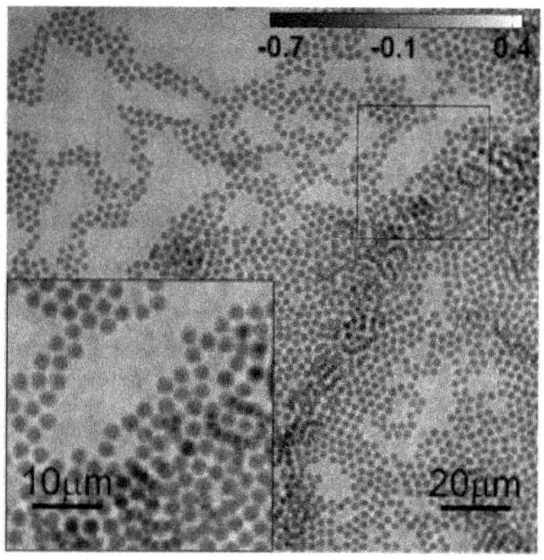

Figure 8.8. A phase contrast image of magnetic colloids (constructed using α in equation (8.2)) measured using coherent x-ray radiation at the Diamond synchrotron. Specifically, ptychography was used, a method in which the phases are calculated using the relative positions of overlapping coherent diffraction patterns, by displacing the beam relative to the sample with known offsets. Reprinted with permission from Edo *et al* 2013. Copyright 2013 by the American Physical Society.

8.3.6 Coherent x-ray imaging

The refractive index of a sample is the phenomenon that modulates the contrast observed in conventional microscopy experiments and it is also an important issue with x-ray imaging. It is often less well recognized that the refractive index has two independent components, the absorptive component (β) and the real component (α)

$$n(\omega) = \alpha(\omega) + i\beta(\omega). \tag{8.2}$$

Furthermore, each of these components depends on the frequency of the photons incident on the material (ω) (Waigh and Rau 2012). Thus, contrast, in an experiment can be either due to the real or the complex refractive index and it depends on the frequency of the photons used (x-ray contrast is different from visible light contrast, which is in turn different from infra-red contrast).

Figure 8.8 shows an example of a lens-less coherent x-ray image used to investigate magnetic colloids, which uses phase contrast, i.e. α in equation (8.2). Similar techniques are possible with biological specimens, although these medical uses have been to date fairly limited due to the prohibitive costs involved to make coherent x-ray sources and the conservative nature of radiation safety procedures that are commonly used, i.e. people are slow to adopt new techniques in medicine that use potentially harmful ionizing radiation.

References

Buxton R B 2009 *Introduction to Iunctional Magnetic Resonance Imaging: Principles and Techniques* (Cambridge: Cambridge University Press)

Callaghan P T 1993 *Principles of Nuclear Magnetic Resonance Microscopy* (Oxford: Oxford University Press)

Drexler W and Fujimoto J G 2015 *Optical Coherence Tomography: Technology and Applications* (Berlin: Springer)

Edo T B, Batey D J, Maiden A M, Rau C, Wagner U, Pesic Z D, Waigh T A and Rodenburg J M 2013 Sampling in x-ray ptychography *Phys. Rev.* A **87** 53850

Ferraro J R 2002 *Introductory Raman Spectroscopy* (Edinburgh: Academic Press)

Georgiades P, Allan V J, Dickinson M and Waigh T A 2016 Reduction of coherent artefacts in super-resolution fluorescence localisation microscopy *J. Microsc.* **264** 375–83

Jaradat S, Harvey M and Waigh T A 2012 Shear-banding in polyacrylamide solutions revealed via optical coherence tomography velocimetry *Soft Matter* **8** 11677

Masters B R and So P T C 2008 *Handbook of Biomedical Non-linear Optical Microscopy* (Oxford: Oxford University Press).

Mertz J 2010 *Introduction to Optical Microscopy* (Denver, CO: Roberts).

Rust M J, Bates M and Zhang X 2006 Sub-diffraction limit imaging by stochastic optical reconstruction microscopy (STORM) *Nature Methods* **3** 793-5

Waigh T A and Rau C 2012 X-ray and neutron imaging with colloids *Current Opinion in Colloid and Interface Science* **17** 13–22

Chapter 9

The structure of carbohydrates—the perfect chip

9.1 Introduction

Carbohydrates tend to have a Cinderella status in molecular biology, compared with other more popular glamorous molecules, such as proteins, nucleic acids and lipids. The more popular molecules tend to attract more funding and are targeted by the most successful researchers. This is more due to technical challenges in understanding the structures of carbohydrate molecules, rather than any lack of important phenomena in living organisms associated with them. There are thus many problems to attract the attention of physicists.

Carbohydrates fulfil lots of interesting functions. They play crucial roles in foods (mankind's main source of energy; ask a *Cordon Bleu* chef about their favourite polysaccharides and expect a long conversation), biomaterials (textiles, building components, paper etc), and the extracellular matrix (e.g. a vital component of cartilage in knee joints). In addition, they form adhesive structures between cells, protect bacteria against antibiotics (in the form of both biofilms and capsules) and are a key player in immunosignalling; among their many other duties. Carbohydrates are thus relevant to a significant number of billion dollar industries and a huge range of human diseases. Clearly these molecules are undeniably of crucial importance to mankind and, as an integral part of our bodies, people cannot live without them. What are the stumbling blocks associated with research into carbohydrates?

Most carbohydrates are semi-crystalline in concentrated forms, such as the wooden desk the author is currently sat in front of or the starch he will eat for his evening meal, and these materials contain sections of amorphous material interspersed with crystalline blocklets. This seriously complicates the process of structural elucidation of carbohydrates with x-ray techniques, which are a standard high resolution tool in molecular biology. The crystalline blocklets are oriented at random inside the solid, which blurs the diffraction patterns measured in x-ray experiments. Blurred diffraction patterns are hard to interpret to create structural

doi:10.1088/978-0-7503-1374-2ch9 © IOP Publishing Ltd 2017

models. Worse still, the amorphous regions are almost completely invisible to the x-ray techniques, so the resultant molecular models lack information on all of these regions.

An even larger challenge is that the basic building blocks of carbohydrates appear in many more flavours than are possible with other giant biomolecules, such as proteins and nucleic acids. There are 10 main primitive possibilities in humans for carbohydrate monomers, but a huge number of additional chemical modifications can also occur. Worse still there are often as many as six ways to glue two building blocks together in carbohydrates, compared with proteins and nucleic acids that only have a single possibility. Chair and boat conformations are also both possible for carbohydrate monomers that commonly self-associate into small circles, which provide additional surprises for structures at the Angstrom level of resolution. Even worse, carbohydrates are predominantly branched, with multiple types of connection demonstrated simultaneously for the same basic building block, which again substantially complicates their topology. In comparison, proteins and nucleic acids tend to be linear as a rule. Cross-links are possible with proteins, such as disulphide bonds, but the peptide bond itself does not branch. Therefore, the relative number of possible combinations of the building blocks in carbohydrate molecules is vast, much bigger than an equivalent protein or nucleic acid chain with the same number of units. This implies a mechanism for high density information storage is possible in carbohydrates (they are indeed found to be important in modulating information transfer in immunology), but to catalogue all the different structural possibilities within these molecules is a logistical nightmare. For example, the high variability of the carbohydrates (in the form of glycoproteins) on the surface of HIV helps it to evade the body's defences (its immune system).

The sequencing of biological macromolecules has had a long and illustrious history in molecular biology, e.g. it famously led to two Nobel prizes for the English chemist Fredrick Sanger, who won one for sequencing proteins (he demonstrated the new methodology with insulin) and the other for sequencing DNA. It seems like an obvious step then, to ask if it is possible to automate the sequencing of carbohydrates. Some good progress has been made in this area, but in general the large-scale sequencing of carbohydrates is still not a reality due to the molecules' extreme degrees of structural complexity. The progress achieved in carbohydrate sequencing still pales in comparison when placed next to the big success stories found with proteins and nucleic acids. Therefore, an obvious (potentially prize winning) question is:

L. *Question*: Create a new method to automate the sequencing of carbohydrates.
Some partially successful solutions to this problem of carbohydrate sequencing are already available. Mass spectrometry and nuclear magnetic resonance (NMR) spectroscopy can be used to determine the structure of small sections of any carbohydrate molecule. However, both these techniques require very expensive apparatus and cannot be easily parallelised for them to be fast enough to handle the sequencing of very large carbohydrate molecules. Arrays of enzymes could be used to chop up large carbohydrate molecules into sufficiently small sections for their

simplified structural elucidation and then statistical techniques could in principle allow the reconstruction of the structure of the original larger molecules. Unfortunately, the techniques with mass spectrometry and NMR are too slow and too expensive to be of large practical use, e.g. to routinely sequence either the carbohydrates that relate to human disease in a hospital or some of the other giant carbohydrate molecules that are important in nature, e.g. to provide quality control of foods. Therefore, one physics based approach to the problem of sequencing would be to invest more effort into the development of higher resolution NMR or mass spectrometry instrumentation.

Specific molecular glues for carbohydrates are known in biology and two standard examples are found within the protein families: *lectins* and *antibodies*. An interesting possibility is that a lectin microarray could be used in tandem with mass spectrometry to sequence carbohydrates. The lectins would identify a range of carbohydrate moieties and subsequent analysis with a mass spectrometer would improve the fidelity of the identification process. However, variable specificity is a problem when both lectins and antibodies bind to carbohydrates. For example, to recognise sialic acids found on the coats of bacteria, lectins can sometimes be sensitive to the exact bonding patterns between neighbouring carbohydrate subunits and sometimes not. The identification of bacterial coats is a major issue with antibiotic resistance. Capsular carbohydrates on the coats of bacteria enable the infections to evade pharmaceutical treatments. Sufficiently specific molecular glues that can attach to capsules are often unavailable, which is unfortunate, since they are needed to target the dangerous ones that need to be eliminated. Similar issues with the need for specific molecular glues occur in the innate immune response of humans; encapsulated bacteria can look identical to human cells to our immune systems and they are thus invisible to these defence systems. Clearly more work is required to understand the molecular processes involved.

An end goal will be to catalogue the whole range of carbohydrate molecules created inside a human cell and how this varies from cell to cell, which is called the *glycobiome* of the cell. The exact sequences of carbohydrates are expected to be well defined for many of their roles inside cells and the information contained in the glycobiome could lead to precious new insights into structure–function relationships and the roles of these carbohydrate molecules in disease.

Additional important structural questions are presented by carbohydrates joined to other molecules, which are also very common in biology, such as glycoproteins (carbohydrates joined to proteins, such as the mucins described in chapter 6), proteoglycans (proteins joined to carbohydrates) and glycolipids (carbohydrates joined to lipids). These mixed molecules can be separated into their constituent carbohydrate, protein and lipid components using enzymes that chop them into bits. Nevertheless, the creation of sequences for the carbohydrate sections continues to be problematic in comparison to the protein and lipid sections, which are much more amenable to standard technologies for sequencing.

The author worked on carbohydrates for his PhD; an experience that took up exactly three years of his life. He submitted his thesis on the final Friday of year three and started a job on the following Monday morning of year four, to begin research on a completely different subject in a different country. At the start of his PhD he would have liked to say he was driven by a burning passion for glucose biochemistry, which provided a guiding light for the choice of his project, and also kept him warm (literally) during the long lonely nights. However, he was trained as a physicist and for the first six months of his project he hated the field of carbohydrates for being horribly complicated (how on Earth could any progress be made?) and that it seemed to be too mundane (his ego became inextricably combined with potatoes, and he lost any hope of intellectual superiority over his fellow man). Over time the subject grew on him and after three years of Pavlovian conditioning (starch combined with research provided a pay cheque) he still unconsciously salivates whenever he reads about a new breakthrough in the field of carbohydrate research. Nevertheless, jokes aside, carbohydrates present an important, relatively unexplored realm of molecular biology, where many new discoveries are yet to be made.

Part of the author's PhD work involved research with minute x-ray beams to probe the structure of starch granules. He tested starches from a wide range of sources, such as bananas, peas, wheats, maize and plantains, but potato starches gave the team some of the best results. X-ray focusing had serendipitously experienced a major improvement in performance at the European Synchrotron Radiation Facility (ESRF) in Grenoble (France) around about the time he started his PhD. The ESRF is a donut shaped particle accelerator for electrons with a circumference of 844 metres. The electrons are wiggled as they move in orbits around the ring, which causes them to release intense beams of x-rays that can be used to measure the structure of molecules with x-ray diffraction techniques. Using the minute (micron-sized) x-ray beams that were newly available, the British team found that helical structures occurred on small length scales in potato starch granules (the helices had a diameter of around one nanometre) that consisted of the side-chains of amylopectin (a giant glucose polymer that is the main constituent of starch) and that these double helices formed on hexagonal lattices. In this research they were neck and neck with a French group, who performed some initial experiments before them, but had a weaker theoretical model, missed some important aspects of hydration on the self-assembly of the structures and were slightly slower to publish. On the next larger length scale (lamellae at nine nanometres), a second set of larger helices also seem to occur in starch, which provides another, as yet unresolved, challenge for structural elucidation. Unfortunately, the lamellar helices are very polydisperse, so experimentally only a single flat line is visible in their diffraction patterns (a layer line, in the jargon of helix crystallography). Possibly it is the first layer line from a lamellar helix, but the scientific jury is still undecided, since there is insufficient information to make an unequivocal decision.

Electron microscopy does provide complementary information on the helical lamellae in starch, but microscopy is often plagued by subtle artefacts and only a single study of helical lamellae exists, uncorroborated by subsequent research, although it is over twenty years old. Cynics say that if an amateur microscopist is told to look for a feature in an image they will invariably find it. The fossilized bacteria that astrophysicists claimed to have been found in a Mars rock in a news conference in 1996 (presented by the US president at the time) is a classic example of this effect. People's brains are very good at searching for patterns in images even when they do not exist. A practical demonstration is to look for clouds that resemble Elvis, if one is found the observers have most likely deluded themselves and the image analysis software in their brain has played a trick upon them.

So here is a carbohydrate question, based on the author's very personal bugbear.

L. *Question:* Determine whether helical lamellar structures do indeed occur in the starch granules found in potatoes.

It is a fairly good bet that the lamellar helices do in fact exist and it is a distinct possibility that they occur in a very broad range of starches (corn, wheat, rice, banana, leaf etc). However, what is the exact morphology of these lamellae (e.g. how are the lamellar helices arranged side by side?) and why do they form these structures in the first place, i.e. what are the physical principles in action? Is there a chiral bending force in effect that causes the lamellae to twist out of the plane due to the well-defined handedness of the molecular helices in starch at the next smaller length scale?

Super-resolution fluorescence microscopy or improved electron microscopy techniques could perhaps be used to revisit the problem of helical lamellae in starch. The single layer line has been reproduced in x-ray diffraction experiments from a number of different research groups, so there is good circumstantial evidence for helical lamellae, but nothing unequivocal yet exists to persuade a sceptical non-believer. New developments in x-ray focusing may also help to solve this problem, since next generation synchrotron sources are expected to provide 10–100 times smaller x-ray beams in the near future. For example, the new ESRF upgrade is expected to provide beams down to the 10 nm size with the prospect that they may soon become diffraction limited (~0.15 nm for hard x-rays). This will open up whole new worlds of possibilities for scanning x-ray diffraction experiments.

Order parameters for continuous phase transitions were famously introduced by the Nobel prize winning Russian theoretical physicist, Lev Landau, to understand the nature of the phases adopted by superconductors and superfluids (liquid helium). He found that much of the fundamental behaviour of the phase transitions in these exotic materials could be described by just using the symmetry requirements of the systems. The theoretical tools he developed were very general and they could then be used by physicists to study phase transitions in soft condensed matter, such as liquid crystals, soaps and polymers. Similar methods were then used by the author to understand the phase transitions in chips (fries for North Americans). The helices in potato starch have a range of possible phase states: they can unwind (a helix to coil transition), they can line up in some direction (to form nematic liquid crystals), they can line up in layers (to form smectic liquid crystals) or they can form glassy phases (in reduced water conditions). It is possible to run through all these different possibilities (and all their different permutations, because they can coexist in heterogeneous polymers) to predict the sequence of phase transitions that will occur during the process of gelatinisation, i.e. the behaviour of starches when they are cooked that results in the creation of chips. At the time the author was very proud of himself in the application of the modern theory of phase transitions to potatoes and, although relatively mathematically unsophisticated, it was unofficially labelled *the big chip theory*. Like all good theories it led to some amusing new predictions (additional thermodynamic phase transitions in potato starch that could be measured using sensitive calorimetry methods) that were later verified by experiment. Specifically, differential scanning calorimetry (DSC) was used to measure the changes in internal energy of starch suspensions during gelatinisation. The experimental verification of the model was followed by a reasonable amount of celebratory dancing by the author, since experience shows there are only a limited number of times that someone actually manages to get it right! A huge variety of other structural phase transitions occur in carbohydrates (e.g. variations in the helical ply structures of celluloses in wood or chitins in the shells of insects) and it is hoped that they too could benefit from quantitative thermodynamic treatments.

Such high technology approaches to the phase behaviour of soft matter might seem a little misplaced when applied to humble starches. Nevertheless, both genetic engineering and conventional breeding strategies with plants can cause large modifications to the molecular structure of starches that are immediately reflected in the ease with which they can be chemically processed. There are thus huge sums of money to be made with starches, combined with huge environmental benefits, for the provision of new insights into the improved efficiency of the processes involved in staling (e.g. food spoilage), acid hydrolysis (e.g. needed for paper making), enzymatic hydrolysis (e.g. needed in beer production and the creation of sugars for soft drinks), and resistant starch (e.g. a major source of dietary roughage).

The list that includes lipids, carbohydrates, proteins and nucleic acids does not exhaust all the possibilities on offer for large molecules in biology (and in particular, plants). Additional possibilities are provided by polyketides, terpenes and polyphenols (lignins in wood are the second most common biopolymer on the Earth, after cellulose, and they are polyphenols). Many of these have similar structural issues as with carbohydrates and clearly there is substantial additional work needed to understand them and they can be very biologically active. In terms of a map of molecular biology, the physical properties of these materials have been little explored. However, a comprehensive understanding of carbohydrates is a useful first step towards a broader, well balanced picture of molecular biology.

Food physics in general is still a very entertaining, active and interesting area for study in the 21st century and carbohydrates represent some of the key questions in the field. A colleague at the University of Leeds made the slightly flippant case for retraining the Umpa Lumpas from Charlie and the Chocolate Factory to equip them with the necessary skill set from theoretical physics, but lots of crucial questions do exist in the field that need some good quantitative research. People have started to worry about food security from the perspectives of over population and climate change. Improved efficiencies in food production could ease many of these worries and provide improved gastronomic delights for everyone the next time they visit a restaurant. Solution to the problems described with carbohydrates in food physics, and the many other questions found in a much wider context, may well allow carbohydrates to give up their Cinderella status and become the Belle of the molecular biology ball.

9.2 Suggested reading

Introductory

Vrachen D V and Weiss G 2013 *Introduction to Bioorganic Chemistry and Chemical Biology* (New York: Garland). An elegant discussion of carbohydrate chemistry that includes a short account of lectin microarrays.

McGee H 2004 *McGee on Food and Cooking: an Encyclopaedia of Kitchen Science, History and Culture* (London: Hodder and Stoughton). Scientific accounts of food have increased in popularity over recent years. This book by McGee is a classic, but there is also Barham P 2001 *The Science of Cooking* (Berlin: Springer) and other recipe based books to consider, e.g. Crosby G 2012 *The Science of Good Cooking: Master 50 Simple Concepts to Enjoy a Lifetime of Success in the Kitchen* (Brookline: America's test kitchen).

Chouffnes D 2016 *Nature's Pharmacopeia: A World of Medical Plants* (New York: Columbia University Press). Tour of some badly understood biological molecules with some of their possible medical applications.

Search on the web for the 'Harvard cooking and science lecture series'. The lectures follow a 'chef meets physicist' format. The Harvard physicist introduces a phenomenon in soft matter physics and the chef then explains how to use it in a recipe, e.g. to create a soufflé.

Advanced

Waigh T A, Donald A M, Heidelbach F, Reikel C and Gidley M G 1999 Analysis of the native structure of starch granules with small angle x-ray microfocus scattering *Biopolymers* **49** 91–105. The author's PhD work on the lamellar helices that as likely as not do occur in potato starch. The evidence presented is from x-ray scattering experiments performed at the ESRF in Grenoble.

Waigh T A, Gidley M G, Komanshek B U and Donald A M 2000 The phase transformations in starch during gelatinisation: a liquid crystalline approach *Carbohydrate Research* **328** 165–76 Describes elements of the Big Chip Theory to understand the gelatinisation of starches.

9.3 Technical details

9.3.1 Gallery of carbohydrates

The 10 most important monomers found in human oligosaccharides are β-D glucose, β-D galactose, α-D mannose, β-D glucuronic acid, β-N-acetyl-D-glucosamine, β-N-acteyl-D-galactosamine, β-D-xylose, β-L-fucose, β-N-actely-D-neuraminic acid and β-D-ribose. This is a long list, but not so long that it appears an impossible task to sequence carbohydrate polymers that contain mixtures of these ten basic units (Vrachen and Weiss 2013). If all the carbohydrates had the same linkage, a carbohydrate chain containing N links would have 10^N combinations. However, inclusion of six different possible linkages increases this number to $\sim 10^N.6^{N-1}$ which is a much larger number for $N > 2$ than 4^N for nucleic acids or 20^N for proteins.

Furthermore, the calculation for carbohydrates is a lower limit, since many more types of monomer may be involved (lots of chemical modifications are possible, such as sulfonation). Clearly this huge number of possible combinations provides a large headache for would-be sequencers and we have not yet started to discuss additional complications associated with branching, e.g. important changes in topology can occur, such as ring molecules.

9.3.2 Boat and chair conformations

To add to the structural complications with carbohydrates, there are two possible forms for each sugar monomer called (a) the boat and (b) the chair conformation (figure 9.1) (Vrachen and Weiss 2013).

9.3.3 Blurred diffraction patterns

As an example of a blurred x-ray diffraction pattern from a semi-crystalline carbohydrate, see figure 9.2, which shows a diffraction pattern from a 10 μm region of a single potato starch granule (Waigh *et al* 1997). The data is in good agreement with a model for crystals of helical amylose (a glucose polymer, closely related to the main component of starch, amylopectin). Nevertheless, in general there is insufficient information from fibre diffraction patterns to create molecular models *ab initio* without a great deal of complementary information. For example, the Watson and Crick model for double helical DNA was corroborated by fibre diffraction patterns (from Rosalind Franklin), but could not have been deduced from them alone. The model needed extensive additional molecular information to restrict the number of choices.

9.3.4 Model of helical lamellae in potatoes

On the next largest length scale in potato starch, beyond the crystalline structures shown in figure 9.2, a helical structure of the amylopectin molecular helices

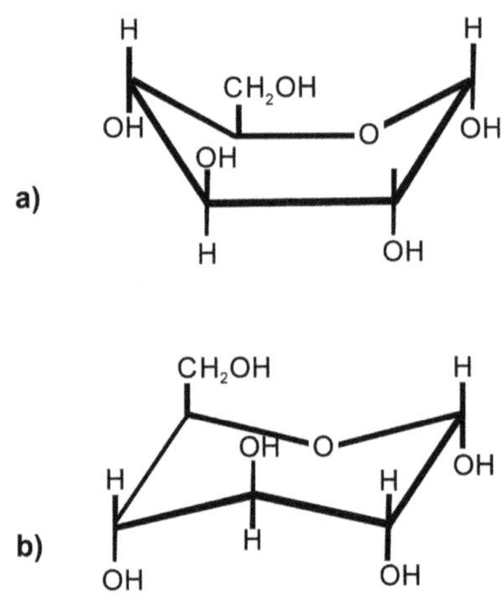

Figure 9.1. (a) The boat and (b) the chair conformations of a glucose monomer.

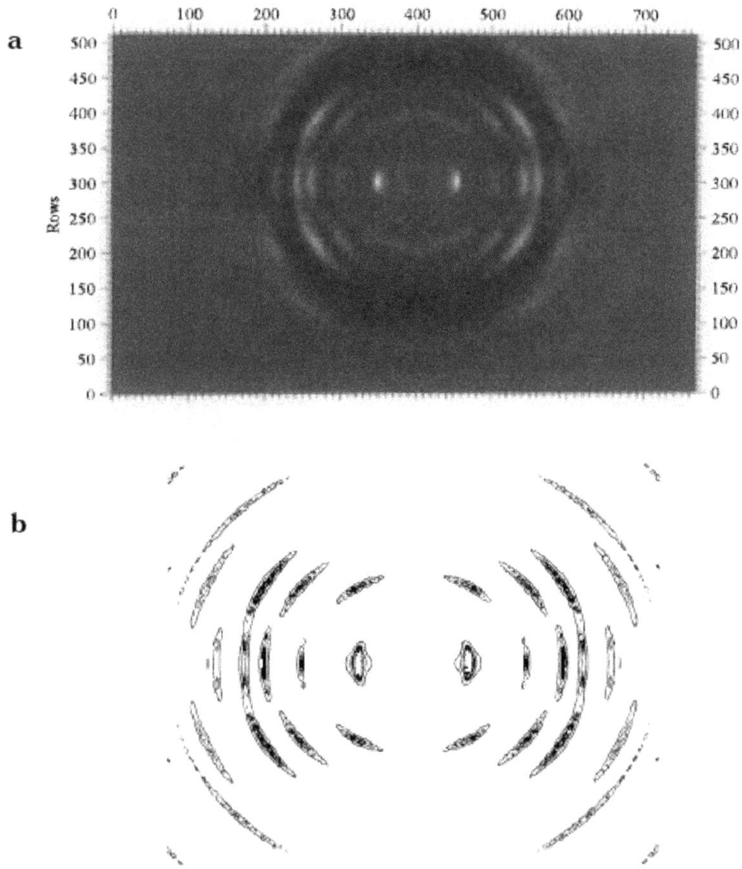

Figure 9.2. Comparison of fibre diffraction patterns from a microfocus x-ray experiment. (a) Averaged microdiffraction pattern from the edge of wet potato granules. (b) Diffraction pattern calculated using the Cerius software package with a unit cell for an amylose double helix. Reprinted with permission from Waigh *et al* 1997. Copyright 1997 American Chemical Society.

(figure 9.3) is also thought to occur. However, here there is only a single layer line in the diffraction pattern and there is thus even less information for building models than at the smaller length scales (figure 9.2) (Waigh *et al* 1999).

9.3.5 Phase transitions in food

The chapter's subtitle is 'the perfect chip'. Creation of chips involves the heating of potatoes in the presence of water and scientifically the process is called *gelatinisation*. To create the perfect chip the phase transitions during gelatinisation need to be carefully controlled. Inspired by Landau's theory for phase transitions, the process of gelatinisation in amylopectin can be monitored using three order parameters (figure 9.4), the helicity (h), the nematic order parameter (ϕ) and the smectic order parameter (ψ) (Waigh *et al* 2000). The choice of the last two order parameters was inspired by studies into liquid crystalline materials, e.g. the type of molecules found

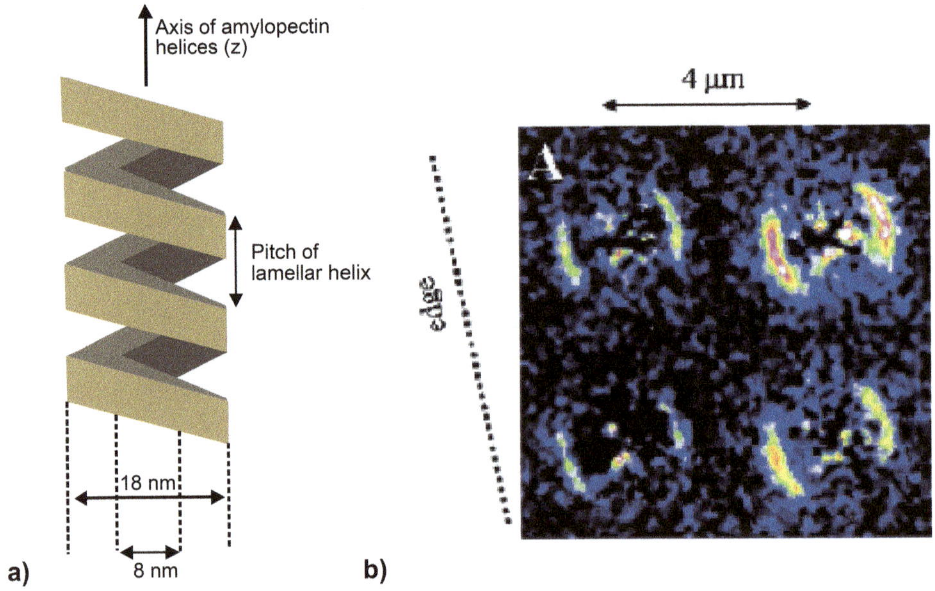

Figure 9.3. Lamellar helices are thought to occur in potato starch (Waigh *et al* 1999). (a) An individual lamellar helix and (b) a diffraction pattern rotated by ~90° from a microfocus x-ray experiment from the lamellae at the edge of a potato starch granule. Reprinted with permission from Gebhardt *et al* 2007. Copyright 2007 American Chemical Society.

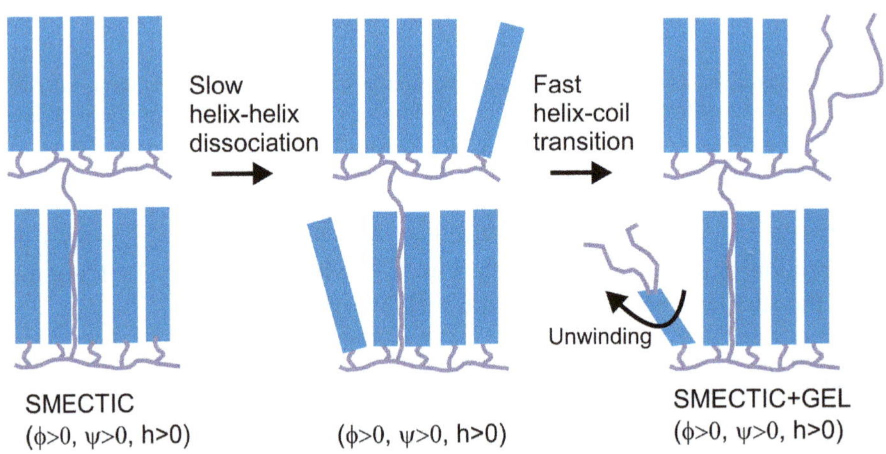

Figure 9.4. A two-stage process is thought to occur in the gelatinisation of starch in excess water. The first stage involves a slow dissociation of the amylopectin helices side by side. Subsequently, a fast helix-coil transition occurs. ϕ is the orientational order parameter, ψ is the lamellar order parameter and h is the helical order parameter. Reprinted from Waigh *et al* 2000. Copyright 2000, with permission from Elsevier.

in computer display screens (Collings and Hird 1997). The amount of water in the starch suspension is one of the most critical parameters that determine the phase behaviour, since it drives the self-assembly of the lamellar structures.

9.3.6 Glycoproteins

Due to their chemical heterogeneity (primarily the carbohydrate regions), glycoproteins represent an under explored area of molecular biology and they are discussed in more detail in chapter 6. Figure 9.5(a) illustrates a schematic diagram for MUC5AC (gelling mucin) from the stomach of pigs. Figure 9.5(b) shows a phase diagram for the behaviour of the molecules as a function of pH and mucin concentration. The samples demonstrate pH switchable gelation (i.e. they become gels at low pHs), which is extremely useful in the acidic conditions of the stomach, since the gelled mucin protects the walls of the stomach lining.

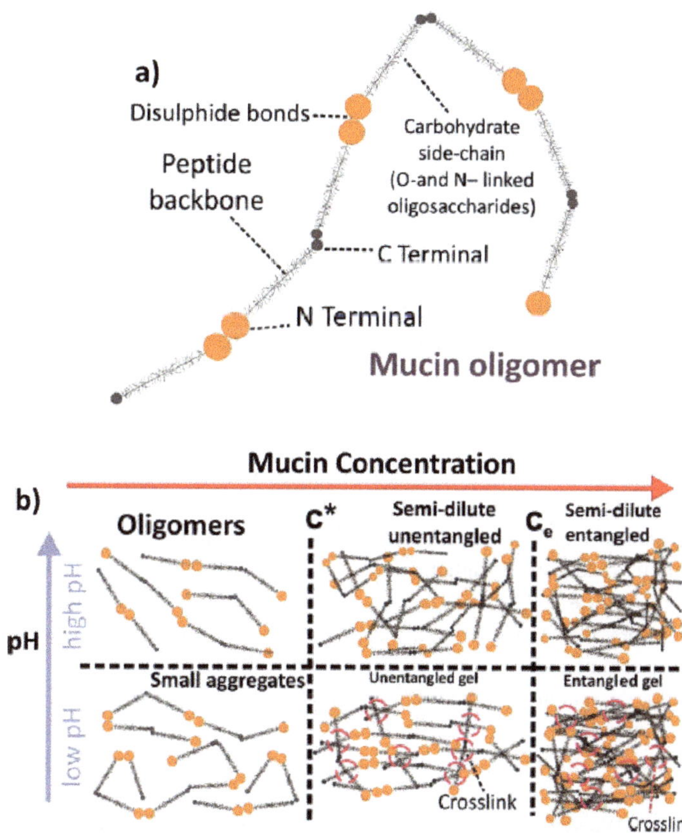

Figure 9.5. (a) A schematic diagram of a mucin oligomer from a porcine stomach. (b) A phase diagram for the different phases of the mucin molecules as a function of pH and mucin concentration (Georgiades *et al* 2014). c^* is the overlap concentration of the molecules and c_e is the entanglement concentration above which the chains start to reptate, i.e. follow snake-like motions along tubes.

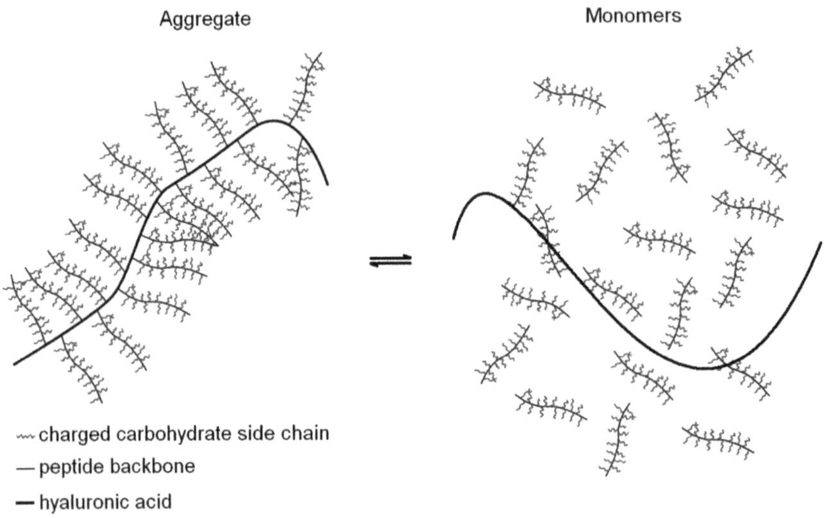

Figure 9.6. Aggrecan is a giant comb polyelectrolyte used in biological shock absorbers, such as cartilage, to dissipate energy. Aggrecan has an unusual bottle-brush of bottle-brushes architecture and can self-assemble from its monomers, if the pH is switched. Reprinted with permission from Papagiannopoulos *et al* 2006. Copyright 2006 American Chemical Society.

9.3.7 Proteoglycans

Aggrecan is a relatively well understood example of a proteoglycan (figure 9.6). Aggrecan has a unique bottle-brush of bottle-brushes morphology and the aggregates can self-assemble as a function of pH from their monomers. Some progress has been made in making a scaling model for its dynamics, but there are still many unanswered questions in terms of its physical properties (chapter 6) (Papagiannopoulos *et al* 2008).

References

Collings P J and Hird M 1997 *Introduction to Liquid Crystals: Chemistry and Physics* (Boca Raton, FL: CRC)

Gebhardt R, Hanfland M, Mezouar M and Riekel C 2007 High-pressure potato starch granule gelatinization: synchrotron radiation micro SAXS/WAXS using a diamond anvil cell *Biomacromolecules* **8** 2092–7

Georgiades P, Pudney P D A, Thornton D J and Waigh T A 2014 Particle tracking microrheology of purified gastrointestinal mucins *Biopolymers* **101** 366–77

Papagiannopoulos A, Waigh T A and Hardingham T E 2008 The viscoelasticity of self-assembled proteoglycan combs *Faraday Discuss.* **139** 337–57

Papagiannopoulos A, Waigh T A, Hardingham T E and Heinrich M 2006 Solution structure and dynamics of aggrecan cartilage *Biomacromol.* **7** 2162–72

Vrachen D V and Weiss G 2013 *Introduction to Bioorganic Chemistry and Chemical Biology* (New York: Garland)

Waigh T A, Donald A M, Heidelbach F, Reikel C and Gidley M G 1999 Analysis of the native structure of starch granules with small angle x-ray microfocus scattering *Biopolymers* **49** 91–105

Waigh T A, Gidley M G, Komanshek B U and Donald A M 2000 The phase transformations in starch during gelatinisation: a liquid crystalline approach *Carbohydrate Res.* **328** 165–76
Waigh T A, Hopkinson I, Donald A M, Butler M F, Heidelbach F and Riekel C 1997 Analysis of the native structure of starch granules with x-ray microfocus diffraction *Macromolecules* **30** 3813–20

Chapter 10

Evolution and antibiotics—Bugmageddon

10.1 Introduction

The World Health Organization has warned for a number of years of the prospects of large-scale antibiotic resistance in clinically important bacteria; 'bugmageddon'. People have misused antibiotics globally for their own ailments, and worse still, indiscriminately used the same ones they depend on to treat their livestock. For example, the misuse of antibiotics with humans involves the failure to complete a full course of a programme of antibiotics to destroy all of an infection, and the use of general antibiotics when more targeted approaches would suffice. Antibiotic resistant bacteria relevant to human health now commonly occur on all the continents on planet Earth. The doomsday scenario is that bacteria will become resistant to all the available antibiotics and medical research will be put back by 100 years. People will begin to die from common infections that up to a few years ago were considered relatively benign; such as *E. coli* in bladders, *Staph* in paper cuts, *Gonorrhoea* in the gonads or *Pneumonia* in the lungs.

An obvious solution is to search for new drugs to kill the bacteria, but this may just slow the development of the problem, rather than provide a long term solution. Bacteria are surprisingly quick to adapt, due to their high genetic plasticity and the fact that many of the drug molecules are already familiar to them (which include many of the natural products used by pharmaceutical companies that are compatible with human physiology) after 3.5 billion years of an evolutionary arms race with their neighbouring organisms. Drugs, such as penicillin, are naturally used by fungi as part of their evolutionary arms race against bacteria. The advantage of these antibiotics is that they are very well targeted against the bacteria, but unfortunately they are also very familiar to the bacteria, which have consequently developed efficient contingency plans. Can the game of antibiotic resistance be played faster and more skilfully than the bugs?

doi:10.1088/978-0-7503-1374-2ch10 © IOP Publishing Ltd 2017

L. *Question:* Create a detailed picture of the games that are played by medically relevant bugs to confer on them antibiotic resistance.

The word *game* may seem a little misplaced in this context, particularly with the emphasis on medicine and fatal infections, but reference has been made to a brand of mathematics originally developed to describe economics; *game theory*. With bacteria the size of the populations of the bugs can be substituted for monetary wealth in the game theories. The evolutionary game is therefore played by the bugs to optimise the number of individuals. Currently there are thought to be around 5×10^{30} bacteria that exist on planet Earth; a fantastically large number. This huge number is on the order of the diameter of the observable universe measured in millimetres, i.e. 8.8×10^{29} mm. It is also the number of *E. coli* (approximately one cubic micrometre in volume) that exists in 10 cubic kilometres, which seems a reasonable estimate (a quick sanity check). Alternatively, it corresponds to 50 tonnes of bacteria for every human on planet Earth.

Only 10 000 bacterial species have so far been rigorously described by biologists, compared with a crude estimate of a total of 10^{19} species that are thought to exist. A huge range of behaviours will be demonstrated by this vast number of bacteria, so the discussion needs to be restricted to standard medically relevant strains that associate themselves with humans, to be more pragmatic about the prospects for advances in the short term. Thus, order of magnitude estimates show that to date less than 10^{-17} percent of the bacteria in existence have been explored to any extent. Clearly this is a much smaller number than that implied by the very conservative limit of ‘less than 1%’ often quoted in microbiology textbooks.

In addition to the exorbitant amount of time and resources that would be required to explore the huge amount of bacterial species that exist on planet Earth, it is our inability to grow most of them in a test tube that hampers the research. Without the ability to culture the majority of bacteria, it is very difficult to study most bacterial microorganisms in any detail.

Just restricting the discussion to bacterial populations on a single human being there are found to be around 10^{11} bacterial cells (mostly inside people’s intestines), which is a factor of 10 times larger than the number of human cells in a single person (due in part to the smaller size of bacterial cells). This factor of 10 is much quoted in the microbiology literature, but was recently downsized to around 1.3 by Sender *et al* bioRxiv, 2016. However, there are still a huge number of bacteria associated with a single human body according to whichever calculation you choose to believe and the bugs are in the majority. This huge personalized selection of bacterial fauna can have a large impact on an individual’s health and wellbeing. It is thus possible to consider a human as a super-organism comprised of human cells interacting with a diverse ecosystem of microbes (around 10 000 separate microbial species).

An obvious extension to the discussion of game theory with bacteria is to decide whether all of the bugs’ strategies can be exhausted in the game of antibiotic resistance. Bugs often have a series of back up contingency plans if they are confronted with a potential threat. Of course the bugs are not sentient, but 3.5 billion years of warfare with the other inhabitants of planet Earth has led to the evolution of numerous robust survival strategies. The default Plan A for virtually every bacterium is to make as many copies of itself as possible and this typically results in the exponential growth of its population. Often the bugs hedge their bets in this respect. The majority of individuals try to reproduce as fast as possible, but a few persistor cells follow slightly less risky lifestyle choices, where they reproduce less quickly, and this buffers the population against extinction in the long term.

One *E. coli* (a fast dividing bacterium found in human guts) will divide into a number of *E. coli* that is equivalent to the mass of an aircraft carrier after a single day and the mass of planet Earth after two days, provided that their growth is not limited by the available nutrients. When the growth rate of the bacteria is inevitably limited by environmental conditions, such as the exhaustion of their food supply or antibiotics specifically targeted at them (often released during co-competition with other microorganisms), they need to have some contingency plans. Frequently these contingency plans are multi-layered, so there is not only a Plan B, but a Plan C, D, E etc, as well.

For example, antibiotic resistant *E. coli* have communities that form capsules, that consist of hairy carbohydrate coats attached to their external lipid membranes. The coats mean there exist survivors if they are treated with standard antibiotics (a possible Plan B), since the antibiotics cannot penetrate the capsule walls. The coats also tend to help stop the bacteria becoming meals for other cells, e.g. they inhibit phagocytosis. If an antibiotic was created to penetrate these hairy coats, then their plan C may be to express membrane proteins that specifically remove the antibiotics from within their cells, i.e. specialised antibiotic pumps made from newly expressed membrane proteins. Another successful strategy (plan D, say) is for the bacteria to embed themselves in a thick viscoelastic biofilm, so antibiotics are excluded from their environments, since the antibiotic molecules cannot easily travel through the fine polymeric mesh of the biofilm. This is a very common strategy and investigation of the occurrence of slime in people's daily life finds a large proportion of it is associated with microbial biofilms, be they mildew on their shower curtains, scum on their tropical fish tanks or plaque on their teeth. The obstructed motility mechanism to antibiotics that is often used to explain the occurrence of biofilms is slightly contentious in the field, since the antibiotics are often much smaller than the mesh size of the biofilm, i.e. the antibiotics should be able to diffuse through the biofilm relatively unperturbed. Something more subtle may be going on, such as the coherent activity of the whole bacterial population inside the biofilm to exclude antibiotics from their environment, e.g. making sealant molecules to protect the edges of the biofilm, adsorption of the antibiotics onto the polymers in the biofilm or members of the community sacrificing themselves for the good of their neighbours (although this is just speculation at this stage).

Additional strategies adopted by bacteria against antibiotics include the modification of pores on their surfaces so that antibiotics cannot pass through them, the release of denaturing enzymes that cut up the antibiotics when they enter a cell, the modification of the molecular targets for antibiotics inside a cell so they are no longer recognised by the antibiotics (stealth molecules) and switching their metabolic levels to a very slow state (the persistor cells referred to previously, since antibiotics often only affect actively growing cells).

The design plans for these survival strategies are sometimes written on the DNA of a bacterium that occurs in its chromosome (the main storage unit of its DNA), but they can also occur on extra DNA upgrades that are included on *plasmids* (additional separate circular sections of DNA that float around inside bacterial cells). Plasmids can be easily transferred from one bacterium to another, and once one has a successful plasmid upgrade to counteract an antibiotic, it can very quickly be shared among all of its neighbours. Indeed, the genetic basis that defines a specific species of bacteria is on a slightly shaky foundation, because bacteria find it so easy to mix and match their DNA chains. The genomes of different individuals from a species of bacteria tend to match their genes at the ninety percent level of similarity (say one *E. coli* with another *E. coli*), whereas mice resemble humans by 92%, chimpanzees resemble humans by 98% and humans resemble other humans by 100%. Clearly there is a huge variation in the genomes within a single bacterial species. Bacteria thus find it fairly easy to pick up features from other bugs that are in principle a separate species.

Bacteria and fungi release antimicrobials to kill other species that are competing for their resources. Nevertheless, they need to be immune to the molecules they themselves create. Thus, there is a reservoir of protection strategies written into their DNA, which can occasionally be passed on to other species. Many of the antibiotics used by man are from natural sources (e.g. penicillin from fungi) and the bacteria already possess strategies to combat them.

The challenge is therefore to exhaust all of the bugs' range of strategies, which do not necessarily occur in the simple consecutive order, A, B, C, and D, that were used to introduce them. Mankind could consequently checkmate the bugs every time, to limit their growth and eliminate the more dangerous varieties.

Ideally to protect everyone against bacterial infections, people need to know the bugs' strategies before the game starts (when an infection is acquired), to provide well-informed medical interventions that are applied in a timely manner. This is where realistic research programmes are required, combined with accurate mathematical models. Ideally, strategies can also then be developed that are targeted to each type of specific bug, rather than the use of broadband strategies.

Overuse of broadband antibiotics is a major problem in the war against infections. Many microorganisms are harmless to their hosts and some can even be beneficial, e.g. bacteria in human intestines can synthesise the vitamin B12 that is important for our wellbeing. Other useful jobs performed by gut bacteria in humans include the breakdown of plant fibres in food and the inactivation of toxic substances in other food stuffs. Indiscriminately killing off all the microbial flora in a human body leaves it open to invasion by foreign microbes, since all the hospitable environmental niches continue to exist, and many of the new intruders can be worse than the original inhabitants. Furthermore, over exposure of bacteria to a large amount of an antibiotic increases the chances that they will evolve a biochemical solution to counteract the particular form of chemical warfare that is used against them. Thus, specific antibiotics are needed that only target harmful infections and leave everything else, both the microbial flora that naturally inhabits the human body and the human cells that surround them, completely unharmed.

More subtle games can also be played by bacteria, rather than just layered strategies of A then B then C etc. The varieties of game that are played become particularly interesting in mixed populations of microorganisms. For example, three different varieties of bug can play the equivalent of a rock, paper, scissors game amongst themselves (which can be extended to a game of rock, paper, scissors, Spock for four types of bug; with N fold generalizations). Specifically, the first species of bug can inhibit the growth of a second species, which inhibits the growth of a third species, which in turn inhibits the first species, and completes the circle. This can lead to oscillatory growth profiles for the three species. Perhaps more surprisingly, different bacteria species can evolve mutually favourable interactions that are collaborative between themselves, which is an encouraging sign for social scientists that think about the analogous behaviour in humans. However, the underlying Darwinian evolutionary pressure for bacteria always tends to lead to competition as a default activity, and there always needs to be some type of reward in the long term for each species of bug to increase their Darwinian fitness or they will fall foul of the theory of evolution.

Bacteria can alternate between a range of different strategies during their lives and mixtures of strategies are often demonstrated simultaneously. Game theory is a rich source of hypotheses to describe bacterial strategies. Another useful concept it provides is the Red Queen hypothesis (from *Alice in Wonderland*) in which the Queen tells Alice 'It takes all the running you can do to keep in the same place'. All organisms need to work hard to keep up with a constant evolutionary arms race for resources or they will quickly become extinct. Inflexibility in the bugs' responses to these evolutionary pressures will deliver a death knell for most types of microorganism.

For a more specific example, consider *E. coli* in urinary tract infections (UTIs). UTIs most commonly occur in women, where they often cause substantial discomfort, but less frequently they can become fatal if the bacteria pass into the blood stream. The gender based differences in UTIs are thought to be due to a hydrodynamic effect, since urine is discharged at a higher pressure in men than in women. The *E. coli* in UTIs are covered in a carbohydrate capsule. These capsules protect the bugs from antibiotics and cause a marked increase in their virulence. Lots of varieties of capsule exist for *E. coli*, and the analogy is made with a packet of multi-coloured M&Ms. One type of capsule flavour (the carbohydrate coating) occurs in bacteria that cause UTIs and another one occurs in those that cause meningitis, but all the bacteria housed within the different coatings are very similar forms of *E. coli*. Each type of coat chemistry can cause a different disease and hundreds of varieties currently exist with *E. coli*, e.g. the blue ones could cause meningitis, the red ones UTIs and the green ones gastroenteritis. The word *capsule* is a misnomer in this context, since the coats are not solid shells and they cannot resist a shear stress. Capsules are actually charge and sterically stabilized polymer brushes (furry charged coats), that help repel antibiotics or other marauding cells that might want to eat the bacteria. The charge and steric stabilisation of nanoparticles is commonly used by man in chemical engineering, e.g. it finds applications in paints, but bacteria created this mechanism many millions of years before man rediscovered it. Nevertheless, *capsule* is the current scientific jargon for the bacterial brushes, and at least for now, the field is stuck with it. Specifically, in humans, capsules help to repel lymphocytes, such as white blood cells, that are used in people's immune response systems to combat infections.

What games are played by *E. coli* in urinary tract infections (UTIs)? Not all the bacteria express capsules with *E. coli*, since those without the coats have an advantage during the infection of new cells (they adhere more easily to the walls of the urinary tract), whereas those with the coats are more antibiotic resistant. The bugs appear to be playing a game of bet hedging. After infection, when the bacteria are inside the human cells, they regrow their coats, since they are a useful protection against intracellular defence mechanisms and cellular adhesion is no longer a necessary part of their life cycle.

Antibiotics have been around for millions of years (probably even billions of years) and are produced as part of the chemical arms race between different organisms, so bacteria have experienced a strong evolutionary pressure to create effective defences against them and *E. coli* have thus been quick to develop antibiotic resistance in humans. Even bacteria themselves often produce antibiotics to kill other rival bacteria, which helps them lever more resources from a particular environmental niche. Antibiotic secreting bacteria need to be resistant to the antibiotics they produce to kill their rivals, which provides a reservoir of antibiotic resistant genes that can be transferred between microorganisms.

The structure and function of capsules is clearly a key issue in the virulence of *E. coli*.

L. *Question:* Deduce the molecular structure of an *E. coli* capsule and provide a quantitative explanation of how it depends on the physical environment (pH, salt, temperature etc). Also determine the molecular machinery that creates capsules in *E. coli*.

To solve this problem will require the combination of some carefully performed genetic techniques with some methods for high resolution microscopy, e.g. super-resolution fluorescence microscopy. Polymer physics models exist for the forces between polymer brushes calculated on the basis of the statistics of the individual polymer chains and these could relatively easily be adapted to describe the forces between capsules.

Evolution can quickly lead to differentiation in bacteria (the creation of new species), since they often divide very quickly, so there is scope for high throughput genetic studies to map interrelationships between separate varieties of bug that compete with one another. For example, different species of *Pseudomonas aeruginosa* are found in different environmental niches in the lungs of patients with cystic fibrosis. The current assumption is that the speciation of the bacteria occurs during the patients' life times, due to the different environmental pressures in different parts of the lung. Quantitative models of this behaviour would be very useful.

In addition to capsules, bacteria can create large protective structures called *biofilms*. The oldest fossilized evidence for life forms on planet Earth are from biomats (biofilms formed by oceanic bacteria) that are over three billion years old. It is thus possible that naturally evolved antibiotics are of a similar age. In the modern world the majority of naturally occurring bacteria still spend the majority of their lives embedded in the biofilms. Biofilms are in general improperly understood, but they need to be disrupted in cleaning strategies to remove infections and they are therefore of crucial importance with clinically important bacteria. The biofilms are thus separate targets for clever chemistry, in addition to the bacteria themselves, and a separation is made between anti-biofilm and anti-bacterial targeted drugs. A better understanding of biofilm structure and function should lead to better informed anti-biofilm treatments, e.g. molecules in garlic and tea are thought to be very effective in disrupting biofilms, which may explain why they are created by the plants in the first place.

L. *Question:* Make an accurate quantitative model for the creation and regulation of the structure of bacterial biofilms, e.g. with *Staphylococcus aureus* which is an important type of bacteria that causes infections that are acquired in hospitals.

For example, the forces that determine the initial adhesion events for the first colonisation events on surfaces need to be understood. They will be a combination of van der Waals forces, screened electrostatics and hydrogen bonding, but are expected to be sensitively dependent on the geometry of the bacteria. Furthermore, bacteria have an arsenal of specific adhesion molecules, the molecular equivalents of grappling hooks, and the forces on these entities need to be quantified. Unusual fundamental aspects of adhesion forces at small length scales also need to be considered, such as the adhesion force measured depends on the time scale at which it is measure (described by Bell's equation). If the initial adhesion events could be prevented, so too could the colonisation of the surface by the bacteria with the subsequent creation of potentially life-threatening biofilms, e.g. a simple sugar (mannose) was recently found to disrupt adhesion forces in disease causing *E. coli.*

MRSA, Methionine Resistant Staphylococcus Aureus, is a strain of bacterium that is named after the antibiotic to which the *Staphylococcus Aureus* (Staph) bacteria are resistant. MRSA often makes headlines when outbreaks occur in hospitals and they can lead to multiple fatalities. Biofilms are one of the key strategies that MRSA use to evade antibiotics. Staph has a major advantage as an experimental system that the bacteria are spherical (great for modellers) and non-motile. This is not true for bacteria in general and many sophisticated questions in fluid mechanics are presented by the hydrodynamics of swimming bacteria. For example, complex swarming phenomena for the collective motility of bacteria can occur on surfaces during biofilm formation. Bacterial turbulence is also possible (chapters 2 and 5).

Bacteria can communicate with one another using a chemical mechanism called *quorum sensing*. For example, *Pseudomonas aeruginosa* that cause lung infections can signal to neighbouring bacteria when they arrive in a good environmental niche to encourage the collective formation of a biofilm (they all vote on the suitability of the environmental niche and they can then arrive at a quorum to build a new biofilm). The chemical aspects of this process of quorum sensing are relatively well understood. However, a recent surprise has been the discovery of electrical potentials created by bacterial cells (in *Bacillus subtilis*, a fairly benign species that is commonly found in soil) that can also lead to signalling and communication inside biofilms. The process of electrical communication in bacteria results in graded analogue potentials, rather than the digital spikes formed in the mammalian brain (chapter 3). The bacteria are thought to use this strategy of electrical signalling when they run low on metabolites, so the bacteria that live in the outer layers of the biofilm do not starve the inner layers by the exhaustion of scarce resources. The soil bacteria therefore represent a very early form of excitable cell and represent a primitive evolutionary step towards the more sophisticated electrical activity observed in multi-cellular life forms, such as the electrical waves that occur in humans' hearts and brains (chapters 3, 4 and 18). It will represent a major advance if electrical signalling can be demonstrated in other species of bacteria in addition to *Bacillus subtilis* (specifically, a species that is more important to human health).

L. *Question*: Adapt the models for the action potentials found in neurons to quantitatively describe the electrical spikes in single bacterial cells and their communication in biofilms.

Some work has already been done in this area of the electrophysiological modelling of bacteria, but large gaps in the puzzle still exist. For example, only the activity of a single type of potassium ion channel is used in the current models (what are all the other ion channels doing?) and the quantitative agreement with experiment is poor. It is then possible that electrodes placed on a surface could disrupt the process of communication and thus the creation of a biofilm. It is already known that relatively small currents through electrodes can have a strong electricidal effect on solutions of bacteria, although its origin is not well understood.

Other types of microorganism, in addition to bacteria, are also important for disease and many of them also produce biofilms. In general, antimicrobial drugs need to be considered, rather than just antibiotics. For example, in cystic fibrosis both bacteria (e.g. *Pseudomonas aeruginosa*) and yeast (e.g. *Candida albicans* a type of fungus) are significant threats to the patients' wellbeing, when they grow and produce biofilms on the lining of the lung. Antifungal drugs have the added challenge that fungi consist of eukaryotic cells, so their metabolism is very similar to human cells. This makes the creation of well targeted drugs much more challenging against fungi, since they are more likely to have adverse effects on the human cells that surround them. Many of the drugs currently used are natural products, created by fungi to kill rival species, e.g. amphotericin.

Infections created by microorganisms tend to target people with compromised immune systems, such as the elderly, people with cancer or HIV. Antibiotic resistance is also a major issue for people with organ transplants, whose immune responses need to be restricted or the organ will be rejected. Clearly the development of new more effective antimicrobials and strategies for their efficient usage is a huge issue for modern medicine, and it is hoped that biological physics can provide the field with some solid quantitative foundations.

10.2 Suggested reading

Introductory

Davies S 2013 *The Drugs don't Work: A Global Threat* (London: Penguin). An engaging short introduction to modern issues with antimicrobials from the UK's chief medical officer.

Strelkauskas A, Edwards A, Fahnert B, Pryor G and Strelkauskas J 2015 *Microbiology: A Clinical Approach* (New York: Garland) . An up-to-date introduction to many of the implications of microbial research in medicine.

Advanced

Prindle A, Liu J, Ly S, Garcia-Ojalvo J and Suel G M 2015 Ion channels enable electrical communication in bacterial communities *Nature* **527** 59. The discovery of action potentials in bacterial biofilms is described in this article. Action potentials commonly occur in multicellular eukaryotic organisms, e.g. in the hearts and the brains of humans, and they had not previously been seen in such simple microorganisms.

King J E, Owaif H A A, Jia J and Roberts I S 2015 Phenotypic heterogeneity in expression of the K1 polysaccharide capsule of uropathogenic *E.coli* and its down regulation during growth in urine *Infection and Immunity* **83** 2605–13. Bacteria that cause urinary tract infections are found to be more effective in the invasion of human cells when they lack capsules, whereas those bacteria with capsules are more resistant to antibiotics, i.e. the bugs play a mixed strategy of bet hedging.

Lewandowski Z and Beyenal H 2014 *Fundamentals of Biofilm Research* (Boca Raton, FL: CRC). This textbook provides a good overview of physical aspects of the growth of microbial biofilms.

Winans S C and Bassler B L 2008 *Chemical Communication among Bacteria* (Washington: ASM). This book considers some of the biochemical principles involved in quorum sensing with bacteria.

Walsh C and Wencewicz T 2016 *Antibiotics: Challenges, Mechanisms, Opportunities* (Washington: ASM). Detailed academic account of the biochemistry of antibiotic resistance that describes many clinically important strategies that bacteria employ.

Jorth P, Staudinger B J, Wu X and Hisert K B 2015 Regional isolation drives bacterial diversification within cystic fibrosis lungs *Cell Host and Microbe* **18** 307 . Speciation of bacteria occur in different regions in the lungs of patients with cystic fibrosis.

Broom M and Rychtar J 2013 *Game Theoretical Models in Biology* (Boca Raton, FL: CRC) . An advanced mathematical treatment of game theory in biology.

Schmidt-Malan S M, Karau M J, Crede J, Greenwood-Quaintance K E, Brinkman C L, Mandrekar J N and Patel R 2015 Antibiofilm activity of low-amperage continuous and intermittent direct electrical current *Antimicrobial Agents and Chemotherapy* **59** 4610. The electricidal effect of small electrical currents appears to be a robust and relatively universal method to disrupt biofilms. Whether it is related to electrical signalling within biofilms is yet to be proven.

Spaulding C N and Klein R D *et al* 2017 Selective depletion of uropathogenic *E. coli* from the gut by FimH antagonist *Nature* **546** 528–32. Mannose is found to disrupt the adhesion of *E. coli* and thus presents a new mechanism for the treatment of this infection.

Mazza M G 2016 The physics of biofilms: an introduction *J.Phys. D: Appl. Phys.* **49** 203001 An excellent compact account of the physics involved in the creation of biofilms by bacteria.

	Hawk	*Dove*
Hawk	$V - C/2$	V
Dove	0	$V/2$

Figure 10.1. The payoff matrix for the hawk/dove problem from the perspective of game theory (Broom and Rychtar 2013). V is the reward and C is the cost of an injury.

10.3 Technical details

10.3.1 Game theory with bugs

A major challenge with game theory is for it to make fully quantitative predictions to compare with experiments (Broom and Rychtar 2013). To date comparisons tend to be qualitative, but it is a good source of hypotheses to understand underlying behavioural responses of organisms. For example, a classic game used to model biological systems (e.g. to describe fights between stags during the mating season) is called the hawk/dove game. The exact details of the game will not concern use here, suffice to say that each strategy of the interacting organisms can be quantitatively analysed with a pay-off matrix (figure 10.1). Each time two of the organisms have a fight they have four possible outcomes, since each organism has two possible strategies: hawk (aggressive) and dove (unaggressive). The behaviour of the organisms can be predicted based on their tendency to optimize the payoff matrix from multiple fights.

10.3.2 Bacterial population mathematics

Population biology is a large well established field in mathematical biology (Otto and Day 2007). One of the simplest models for bacterial growth can be used to motivate the exponential population growth of *E. coli* discussed in the text. *Malthus' law* is based on a constant division rate (K) for the number of bacterial cells (N) with no death,

$$\frac{\mathrm{d}N}{\mathrm{d}t} = KN. \tag{10.1}$$

A solution to this equation is

$$N(t) = N_0 e^{Kt} \tag{10.2}$$

where N_0 is the initial number of bacteria in a system. Equation (10.2) allows the calculation of the result that a single *E. coli* can divide into a mass equal to the Earth after two days. A more realistic model is given by the *logistic equation*

$$\frac{\mathrm{d}N}{\mathrm{d}t} = r_c N(t)\left(1 - \frac{N(t)}{K}\right) \tag{10.3}$$

where r_c is a constant. This equation predicts a sigmoidal growth curve, i.e. there is a finite carrying capacity of the system for microorganisms, which is a constant at long times.

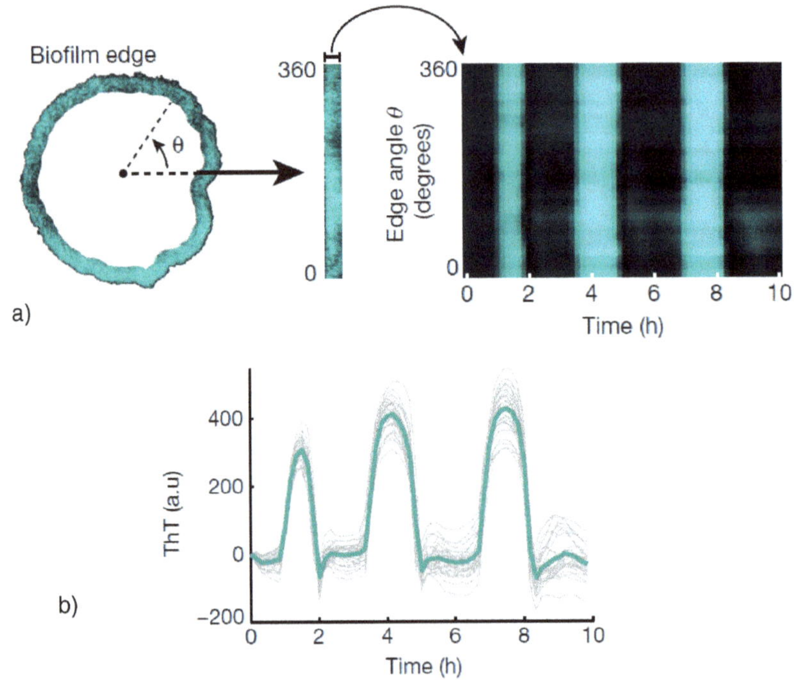

Figure 10.2. Membrane oscillations due to the coherent activity of a bacterial biofilm (*B. subtilis*). The fluorescent dye Thioflavin is used to measure the magnitude of the electrical potentials. (a) A two-dimensional image from fluorescence microscopy of the edge of a biofilm. b) The fluorescent yield as a function of time. The concentration oscillations are fairly slow with a time period of around three hours (Prindle *et al* 2015). Reprinted by permission from Macmillan Publishers Ltd

10.3.3 Electrical communication in biofilms

Chemical communication in biofilms is relatively well established and chemical queues are well known for the creation of biofilms i.e. quorum sensing. Electrical signalling (via potassium channels) has been a much more recent discovery (figure 10.2). We were able to reproduce the effect in our own laboratory using the same strain of *B. subtilis* used by Prindle *et al* so it appears to be a robust phenomenon (Prindle *et al* 2015).

10.3.4 Capsules

Many disease-causing bacteria are coated with long carbohydrate chains called capsules (figure 10.3). Different capsule chemistries are implicated in different diseases and cause different immune responses in humans, e.g. meningitis caused by *E. coli*. There are many open questions regarding their synthesis, e.g. how relatively long lipopolysaccharides (~100 nm) are forced through a hole of a few nanometers in width and tens of nanometers in length, i.e. through an ABC transporter (White and Drummond 2011). There is a large entropic barrier for this motion of lipopolysaccharides (the chains lose conformational entropy when trapped in a pore) and the process is called *translocation*. Translocation is also

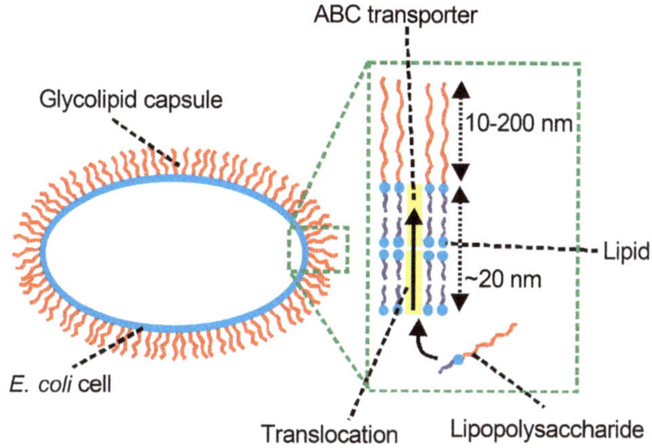

Figure 10.3. A schematic diagram of the capsule on the outside of a bacterium, e.g. *E. coli*. ABC transporters in the membrane are required to move the glycolipids that form the capsule to the membrane surface. Exactly how they do this is not well understood, e.g. there is a large entropy penalty to force a large flexible polymer through a small pore in a membrane.

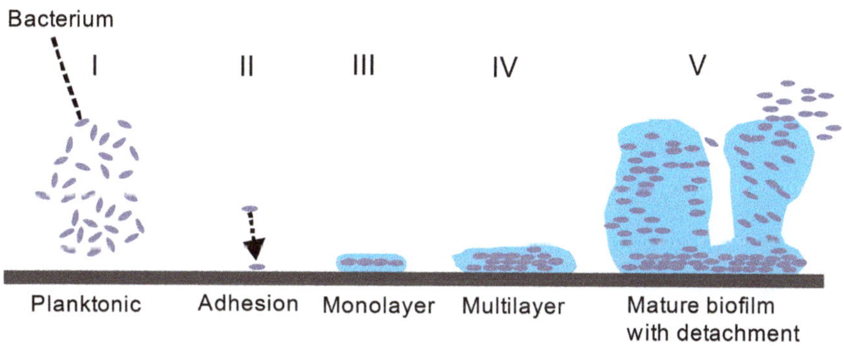

Figure 10.4. Five stages in the growth cycle of a bacterial biofilm: planktonic (I), adhesion (II), monolayer (III), multilayer (IV), and maturation and detachment (V). Bacterial cells are shown in purple and the biofilm is in blue.

observed in single molecule electrophoresis experiments where sequencing is possible by driving a large DNA chain through a small pore using an electric field.

10.3.5 Physics of biofilms

Biologists have found five broad stages that occur during the creation of bacterial biofilms (figure 10.4). These five processes are all determined by physical principles. Bacteria are commonly motile and can swim around in solution (called the planktonic state, *stage I*). When they arrive at a surface adhesion can occur, where the bacteria stick themselves in place using adhesion complexes, e.g. pili (*stage II*). Next, a monolayer forms and the bacteria produce polymeric slimes to bind the film together (*stage III*). Continued growth then occurs, which is accompanied by multi-layer formation (*stage IV*). Finally, mature biofilms form, which can have more complex mushroom architectures (*stage V*). The mushrooms shed planktonic bacteria and these bacteria then swim off to create biofilms in other environmental niches. All the stages in biofilm growth still lack detailed quantitative models, e.g. for the forces in adhesion, the swimming behaviour of bacteria at low Reynolds number, the motility of biochemicals in biofilms etc (Mazza 2016, Lewandowski and Beyenal 2014).

References

Broom M and Rychtar J 2013 *Game Theoretical Models in Biology* (Boca Raton, FL: CRC)

Lewandowski Z and Beyenal H 2014 *Fundamentals of Biofilm Research* (Boca Raton, FL: CRC)

Mazza M G 2016 The physics of biofilms–an introduction *J. Phys. D: Appl. Phys.* **49** 203001

Otto S P and Day T 2007 *A Biologist's Guide to Mathematical Modelling in Ecology and Evolution* (Princeton, NJ: Princeton University Press)

Prindle A, Liu J, Ly S, Garcia-Ojalvo J and Suel G M 2015 Ion channels enable electrical communication in bacterial communities *Nature* **527** 59

White D and Drummond J 2011 *The Physiology and Biochemistry of Prokaryotes* (Oxford: Oxford University Press)

IOP Publishing

Some Critical Questions in Biological Physics
A guided tour around the bugbears
Thomas Andrew Waigh

Chapter 11

The regulation of expression in DNA—huge uncertainties in genetics

11.1 Introduction

The human race now lives in a post-genome world. In 2003 the complete human genome was sequenced, all three billion base pairs of the DNA chains (distributed across 46 chromosomes per cell), with a cost in excess of one billion dollars and a great deal was rightly made of this big successful project. The US president at the time, Bill Clinton, announced the discovery in a special televised news conference. The politicians actually jumped the gun with this announcement in 2000, since it took an additional three years to create a refined corrected version of the genome. Sequencing costs have reduced significantly since this date. The cost of sequencing the complete genome of an individual is now a hundred thousandth of the original cost and it is expected this price will continue to fall, which could soon make the technology accessible to all the world's population. Nevertheless, the mechanism through which all this information is expressed in cells is still not well understood and, without a good understanding of it, genome sequencing will be of relatively limited value. For example, comparative studies can link DNA sequences to disease, but little quantitative understanding can yet be formed of the exact biochemical mechanisms that are in action as a result of variable gene expression. In short, our understanding of genetics is far from complete.

The traditional understanding of how DNA is interpreted inside cells follows the *central dogma of biology*, which is that DNA creates RNA, which in turn creates proteins. Proteins then do many of the important jobs inside cells, e.g. as structural materials or as factories to make other important molecules. Retroviruses provide some counter examples for the behaviour described in the central dogma; since the RNA viruses can cause permanent changes in DNA (they reverse the sequence in the first step of the standard dogma of molecular biology). This presents a major challenge in the treatment of diseases caused by HIV, SIV and leukaemia virus.

doi:10.1088/978-0-7503-1374-2ch11

© IOP Publishing Ltd 2017

However, the central dogma is generally obeyed and it is a very powerful and useful idea, since it implies that changes in an organism's DNA can alter the production of proteins, and it consequently is a cornerstone of the field of genetic engineering. Change the DNA in the correct manner and an organism can then be made to produce lots of useful products. Although very effective, this simple scheme implies genes (stretches of DNA that create a specific protein) are either completely on or off, dependent on the existence of the required stretch along the DNA chain and there is no control of expression. In reality, all biological cells finely regulate the expression of their genes.

The expression of DNA is known to occur in chemical circuits, but their exact connectivity and activity is still largely unknown. The area is consequently hotly contested and it is subject to an intense level of research. Agents in the cells can speed up or slow down the expression of genes or switch them off completely. Exactly how they do this in general still needs to be unravelled, although it is a standard behaviour in all living cells. Analogue calculations can also be conducted by the genes, such as the logical operators AND, NOT and NOR etc that also occur in traditional electronic circuits. Therefore, cells can act as minute computers, which make simple decisions to help coordinate their behaviour in organisms based on their gene circuits. Quantitative numerical modelling is required to describe such behaviour, since even the simplest systems are highly non-linear and models lack simple algebraic solutions. A more detailed exposition of themes in cellular computation can be found in the book *Wetware: a Computer in Every Living Cell* by Denis Bray.

Thus, the regulation of expression of DNA in cells is arguably the most important unsolved issue in genetics and it is of upmost importance in biology. Gene regulation plays an important role in morphogenesis, stem cell research, cancer and nearly every other metabolic event that occurs inside a cell, many of which are critical for an organism's continued existence.

Attempts to reverse engineer the functional design of cells are contained within the field of *systems biology*, and gene circuits are now a traditional theme in this field. Good progress has been made in the understanding of gene circuits within bacteria, but eukaryotic cells, such as those found in humans, are much more complicated and there thus remain many unknowns. Poor gene regulation is implicated in many diseases, in addition to problems directly caused by deleterious mutations that affect the direct products of genes (the amino acid sequences of proteins), such as diabetes, haemophilia or cystic fibrosis.

Messy solutions to metabolic questions are not identical to inefficient solutions in evolutionary terms, but they are not a million miles away. A classic counter example is the optical wiring of the retina in the human eye, which has evolved to be the wrong way around, i.e. the wiring is placed in front of the retina, rather than behind it, so light must be transmitted through the wiring before it is detected (a better solution is used by the octopus, which has the positions reversed). This is a messy inefficient solution that became fixed in a population due to the slow gradual nature of evolution to explore the full landscape of possible solutions to evolutionary challenges. However, in general, inefficient systems can be outcompeted by fitter, more efficient systems during evolution and they tend to be quickly eliminated in a diverse population of an organism. Although inefficiency tends to be eliminated, there is also an evolutionary drive to complicate in some situations, since it can lead to entirely new, game changing behaviours. The ultimate example of the drive to complicate is the evolution of intelligence. A constant stream of new survival strategies can then be created by an intelligent organism that can be game changing for its success, e.g. the development of tools.

As an example of evolution, a mutated membrane protein could become light sensitive, which allows an organism to crudely visualize its surroundings and out-compete neighbouring organisms. In the land of the blind, the organism with a light sensitive membrane protein is therefore king. Over many thousands of generations, through evolution this light sensitive protein can become more and more specialised resulting in the sophisticated organ now seen in man, the eye (that unfortunately has the wiring the wrong way round). Nature thus predominantly tends to create robust, efficient, but often complicated, solutions to problems in the survival of organisms. Nevertheless, when the most up-to-date genetics textbooks are read, the biochemical picture with regard to gene regulation appears like a complicated messy jumble and the mechanisms behind the robust efficient, but complex, solutions to gene control have not yet been clearly elucidated. This is not, in the author's opinion, due to poor scholarship on the part of textbook writers. It is due to the extremely complicated story that nature presents and the fact that it is still an emergent picture with many missing fragments. Therefore, it is currently hard to create a coherent quantitative picture of gene expression that combines all of the separate mechanisms using the standard paradigm of reductionism, i.e. the problem must be broken down into all of its component parts and of all the processes involved separately modelled. Ideally, due to its large size, the hope is that the problem of gene regulation can eventually be broken down into its individual sections and solved on a computer; to create a self-consistent picture for the relative importance of all the different gene regulation mechanisms that work together in tandem. Clearly this can be posed as a research question, another mystery in biophysics that needs to be solved,

L. *Question*: Construct a well-integrated picture of the processes involved in the control of gene expression inside a human cell.

Here a quantitative model is required to convince ourselves that all the competing processes are actually understood. To decide on the tractability of this question on gene expression, some of the immense scope of the problem needs to be first described, to grasp more details of a selection of the mechanisms that nature has evolved to control genes.

First of all consider chromatin; this molecular construct is found in the nucleus of human cells and it is the natural mechanism for the storage of DNA. Chromatin consists of histone beads (compact cylindrical proteins) arranged on a string of DNA that helps it to collapse the string into a manageably small volume, which in turn helps the DNA to fit inside the relatively compact nucleus of a cell. Histones are positively charged, whereas DNA is negatively charged. The two components thus stick together in water to form chromatin due to electrostatic forces. The complexation of DNA with histones takes the form of a phase transition, analogous to more everyday transitions, such as the liquid-to-gas transition in a boiling kettle or the solid-to-liquid transition of the wax in a burning candle. Manipulation of the physical conditions experienced inside the cell (e.g. the temperature, the pH, the osmotic pressure or the salt concentration) can provide insights into how the complexation phase transition is controlled. The total length of DNA in the human genome is around one metre and chromatin is required to compact the DNA, so it is sufficiently small to fit into a nucleus that has a diameter of around five micrometres. However, chromatin (the histone and DNA complex) also provides a framework that determines gene expression, because the tightness of winding of the DNA inside the chromatin can affect the rate of transcription (how information is read off the DNA). In fact the histones appear to have a separate code that controls their interaction with the DNA and thus its tightness of winding. The main varieties of the histone code are acetylation, methylation, ubiquitination and phosphorylation. The specific chemical details are not particularly relevant here. For the purposes of the current discussion, they are just labels for different modifications that affect the winding of DNA. Thus, gene transcription can be controlled through the modification of the histone bobbins to which DNA is attached. Also, variants occur in the amino acids used to construct the histones and they again provide DNA sequences with different tightness of winding and thus different levels of gene expression. The protein bobbins (the histones) therefore provide at least two general mechanisms for the control of genes: chemical modification of the bobbins and the use of different primary amino acid sequences to construct the bobbins in the first place.

DNA famously comes in four different flavours (A, T, C and G) with A paired with T and G with C (Watson and Crick base pairing). Additional minor chemical changes to these four varieties of DNA are also possible, such as methylation, which are carefully conserved through cell divisions. Methylation patterns can switch genes, and other regulatory regions, on or off. Enzymes sensitive to methylation are required during the replication of DNA, if these patterns are to be conserved. During replication, the methylation enzymes shuffle along the initially unpaired DNA strand and methylate the newly synthesised chain to create matched pairs along both of the resultant double helices. This is a simple mechanism to propagate patterns of gene expression in successive cell lines which, for example, ensures that muscle cells produce more muscle cells when they renew themselves in a region of tissue. It is also part of the emergent field of *epigenetics* that describes the inheritance of different patterns of genetic expression from one generation of an organism to the next, that is not just simply encoded in the sequences of the four primitive bases in the DNA (A, T, C and G). Quantitative models are then required to understand how this information is transmitted from cell to cell (chapter 1).

Chromosomal structures at larger length scales in the cell's nucleus also need to be regulated, since they impact on gene expression, and it has already been seen that histones play an important role in this mechanism. Nevertheless, regulation is also in part determined by the DNA sequences themselves. Some DNA sequences encourage the creation of loops, to expose the looped base pairs to transcription enzymes (the molecular machines that interpret them to create RNA), whereas others can act as insulators, and allow the neighbouring sections to create loops in the required manner with limited intra-chain interference e.g. the DNA sequences can provide mechanical isolation that stop the loops overlapping. It is possible to create models for the mechanics of single DNA loops and then test them using modern experimental tools such as fluorescence microscopy, atomic force microscopy and optical tweezers.

In eukaryotes (the types of cell found in humans) 70% of the DNA chains are transcribed into RNA, but only 1% of this RNA codes for genes, i.e. proteins. Much of the other 99% of the RNA that is transcribed plays a regulatory role. Examples include microRNAs (miRNA) that silence other RNA signals, small interfering RNAs (siRNA) which cause messenger RNAs that code for proteins to be degraded and Piwi interacting RNAs (piRNA) that form complexes with Piwi proteins that are important for stem cell type differentiation. Stems cells in all organisms are the initial universal cells that can then differentiate into any other specific type of cell. Stem cells are a current paradigm in regenerative medicine for replacement biomaterials, since they can in principle be plugged into damaged tissue and then cause it to be repaired.

The regulation of genes is often considered from a historical perspective and many of the phenomena involved were first described in simple bacteria, such as *E. coli*, which are found in human intestines (it is easy to obtain new samples). Jacques Monod won a Nobel Prize in 1965 for the discovery of the regulatory gene circuit used by *E. coli* to metabolise lactose, a carbohydrate that occurs in milk. The regulatory systems in bacteria are relatively straightforward; *promoters* bind to DNA and increase gene expression, whereas *repressors* also bind to DNA, but have the opposite effect to promoters and they decrease gene expression. This simple promoters and repressors story is mostly true for bacteria (although regulatory RNA and more sophisticated sigma repressor protein circuits are common occurrences), and also does occur in animal cells, but the complete story in animals is much more complicated, because regulatory RNA plays a much bigger role (miRNA, siRNA and piRNA all take part).

It is thus necessary to consolidate some of these different mechanisms of gene control into a single coherent picture.

L. *Question:* Explain how gene regulation that is based on histones smoothly combines with both the DNA sequence and the RNA mechanisms of gene regulation.

More specifically, quantify the relative importance of each of these three broadly different mechanisms of gene regulation.

Even with bacteria, models for gene regulation are often overly simplified. Following the central dogma of biology, RNA is transcribed from DNA and it is performed by a molecular machine called RNAP (RNAP stands for RNA polymerase). The reaction kinetics of RNA when it is transcribed from DNA by RNAP predominantly occurs in a crowded environment, since cells are packed full of macromolecules with little free space available, which cause sub-diffusive behaviour in the movement of most intracellular particles (chapter 12) that in turn can modify the reaction kinetics. More sophisticated models of diffusion with memory (non-Markovian models) are needed to completely capture this sub-diffusive behaviour. To explore these models experimentally continued advances are needed in microscopy, and fluorescence microscopy is currently the main tool used. Modern improvements in photonic technologies and the development of super-resolution techniques (chapter 8) mean that more accurate physical measurements are now possible than ever before, which can in turn be accurately quantified using new statistical tools combined with high performance computing.

Animals have more sophisticated (eukaryotic) cells than bacteria and in humans there are three molecular machines that can read DNA to create RNA called RNAP I, II and III. RNAP I and III are there to create RNA that has a regulatory role (e.g. miRNA, siRNA and piRNA), whereas RNAP II is needed to create messenger RNA that is intended to be transcribed into proteins, which form gene circuits with other RNA and DNA molecules. Additional RNAPs IV and V can occur in plants, which provide even more complicated modes of control in these organisms. Plants can have genomes that are ten to a hundred times bigger than humans', which is a little surprising from an anthropocentric perspective, although it may just reflect their ability to create a larger arsenal of chemicals to protect themselves from environmental threats.

RNA is constantly degraded inside cells and the degradation rates can be actively varied. There are thus RNA enhancers and silencers inside cells that modify the level of expression of different genes along DNA chains by the modification of the rate at which messenger RNA chains are broken down. RNA polyadenylation and splicing also occur after transcription, which again can change patterns of expression. Exact quantitative expressions for the rates of these RNA reactions are still unavailable in the research field, since relatively little is known about the kinetics of RNA transcription.

It would not be a surprise, if at this point the vast list of strange unpronounceable molecules starts to do a little dance in front of the reader's eyes. A new framework is needed to make sense of all the complexity involved in gene regulation. Perhaps a step backwards should be taken, since there are too many confusing details, and the wood cannot be seen for the proverbial trees.

Therefore, a slightly more nuanced problem will be posed that naturally lends itself to a quantitative analysis and it is key to the understanding of gene regulation.

L. *Question*: Explain how the fine regulation of genes is achieved in a robust fashion by RNA and create an accurate kinetic model for gene regulation by RNA.

There are some very complicated possibilities for statistical models for molecular dynamics in crowded environments, such as fractional reaction-diffusion equations that might be useful to solve this problem of gene regulation by RNA (chapter 12). A type of reaction–diffusion equation was famously used by the English mathematician and code breaker, Alan Turing, to describe pattern formation during morphogenesis, e.g. how the leopard gets its spots. Much simpler models could also be useful to describe the regulation of RNA, as long as they are able to make clear cut predictions that agree reasonably well with experiments, i.e. they produce testable hypotheses.

One of the most promising areas of regenerative medicine is stem cell research. It offers a practical solution to many human diseases. Just plug in replacement cells, as and when they are needed (e.g. to form replacement limbs after a car accident or hair follicles for male pattern baldness), and the majority of issues with faulty biological tissue can in principle be solved. It thus forms a key theme in modern research into regenerative medicine. However, the stem cells need to be differentiated into the correct type of cell to be useful. Differentiation is defined by the pattern of gene expression, e.g. why a muscle cell is different from a kidney cell or a retinal cell. A cocktail of growth factors needs to be added to pluripotent stem cells in an attempt to coax them to differentiate into a required cell type. Often qualitative features of the target cell are captured (e.g. cardiac stem cells adopt the correct shapes and begin to contract in an oscillatory fashion, similar to the behaviour of healthy cells in a beating heart), but the molecular biology of the stem cells is still way off the required target behaviour, e.g. in the chromosomes of the stem cells that have been artificially differentiated the methylation patterns and the histone phosphorylation states are completely wrong, and this means the specific patterns of gene expression will also be wrong. Particularly problematic is the reversion of differentiated stem cells back to their original types once they have been placed inside living people, without the cocktail of growth factors in which they were initially grown, after three to four weeks.

Currently stem cell differentiation is a fairly hit and miss process, and it needs to be made more reproducible if the methodology is to be widely accepted. Much more pure biophysics research into genetic regulation is therefore needed to underpin stem cell research and make the results of these critical medical procedures slightly less haphazard.

Clearly gene regulation is one of the most important unsolved problems in cellular biology. Improved quantitative models could provide a coherent framework to understand the huge number of alternative mechanisms that can play a role in the expression of genes in cells.

11.2 Suggested reading

Introductory

Noble D 2008 *The Music of Life: Biology Beyond Genes* (Oxford: Oxford University Press). A fascinating popular account of the field of systems biology.

Bray D 2011 *Wetware: a Computer in Every Living Cell* (New Haven, CT: Yale University Press). Popular account of the analogue computations performed inside individual cells.

Advanced

Benichou O, Chevalier C, Meyer B and Voituriez R 2011 Facilitated diffusion of proteins on chromatin *Phys. Rev. Lett.* **106 38102**. Primary research that considers some of the physical factors that affect anomalous diffusion of proteins that associate with nuclear chromosomes.

Golding I and Cox E 2006 Physical nature of the bacterial cytoplasm *Phys. Rev. Lett.* **96 98102**. Primary research that demonstrates that the majority of passive motile processes inside bacterial cells are sub-diffusive. This picture of sub-diffusive cellular transport has been robustly reinforced by subsequent research (chapter 12).

Latchman D S 2014 *Gene Control* (New York: Garland). A detailed modern account of the various mechanisms are used that control genes in eukaryotic cells.

Waigh T A 2015 *Physics of Living Processes* (Chichester: Wiley). Describes some simple models for the phase transition that occurs during DNA/histone complexation in chapter 18.

Halford S E and Marko J F 2004 How do the site-specific DNA binding proteins find their targets? *Nucleic Acids Res.* **32** 3040-52. Good pedagogic review of statistical models for the association of proteins with DNA, e.g. regulatory proteins.

11.3 Technical details

11.3.1 Central dogma of biology

The central dogma of biology in its simplest form is shown in figure 11.1 (Alberts *et al* 2014, Latchman 2014). In terms of gene control many embellishments are possible based on this simple scheme. Some of them are shown on figure 11.2.

11.3.2 DNA

The genetic code for the information stored in DNA is found in its base pairs (figure 11.3) (Alberts *et al* 2014). Other modifications are also possible, such as methylation, which plays an important role in cell differentiation, e.g. why a retinal cell is different from a cardiac cell.

11.3.3 Chromosomal structure

DNA needs to be compacted in the nucleus of a cell and cells do this using chromosomes (figure 11.4) (Alberts *et al* 2014). The expression of DNA inside eukaryotic cells (e.g. human cells) is therefore in part determined by its organisation inside the chromosomal structure inside the nucleus. The chromosomes need to be

DNA→RNA→protein

Figure 11.1. The central dogma of biology describes the unidirectional transport of information from DNA molecules to RNA molecules to proteins.

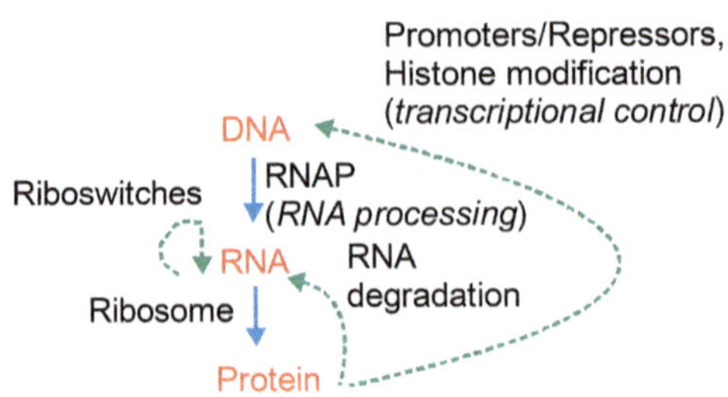

Figure 11.2. Gene control is important in the regulation of the central dogma of biology (figure 11.1).

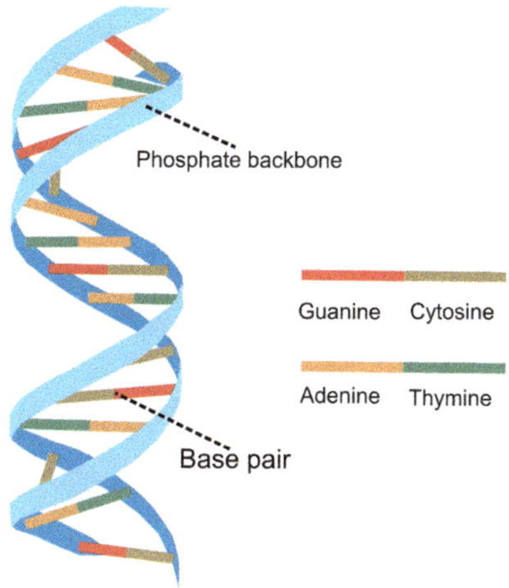

Figure 11.3. The molecular structure of DNA is double helical. The genetic code is found in Watson–Crick base pairing. A is matched with T and G is matched with C.

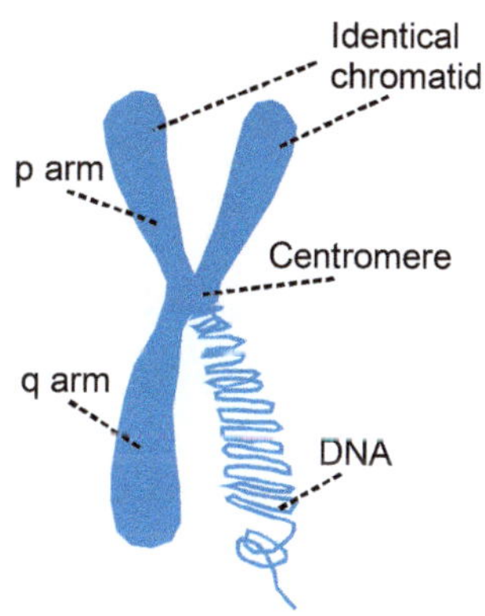

Figure 11.4. Schematic diagram of a chromosome used to store DNA in the nucleus of eukaryotic cells.

less tightly wound if base pairs along the DNA are to be expressed. The condensation of chromosomes from their components can be considered as a phase transition (the first step is complexation of DNA with histone proteins).

11.3.4 RNAP

The protein complex that synthesises RNA from DNA is called RNAP (figure 11.5) (Alberts *et al* 2014). There are three different varieties that occur in humans called RNAP I, II and III. Accurate quantitative models for their kinetics do not yet exist, but single molecule experiments are possible with fluorescence microscopy.

11.3.5 Ribosome

An additional nanomachine is also needed to synthesise proteins from mRNA chains, the *ribosome* (figure 11.6) (Alberts *et al* 2014). The biochemistry of ribosomes

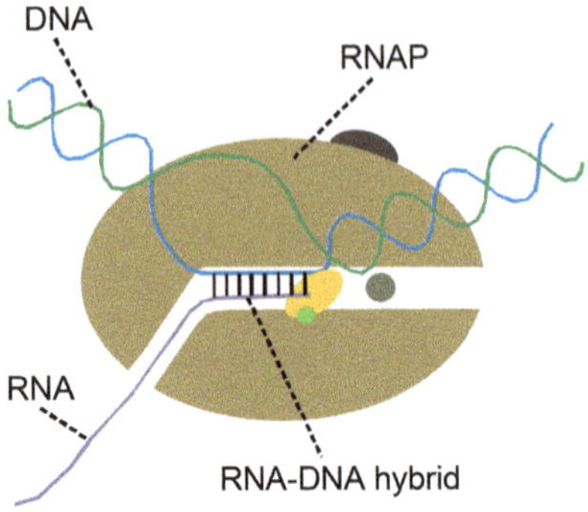

Figure 11.5. RNA polymerase (RNAP) creates RNA from a DNA template. mRNA and regulatory RNA can both be made dependant on the type of RNAP.

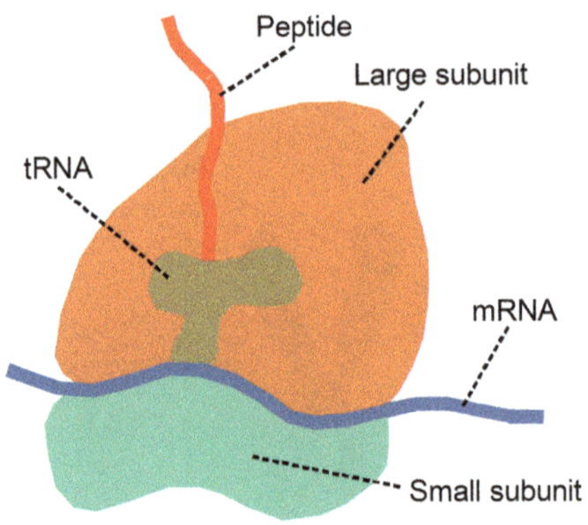

Figure 11.6. Peptide or protein chains can be constructed from a chain of mRNA using the ribosomal construct (the combination of the large and small subunit).

is relatively well determined, but accurate models for the dynamics are still in the early stages of development.

References

Alberts B, Johnson A, Lewis J, Raff M, Roberts K and Walter P 2014 *Molecular Biology of the Cell* (New York: Garland Science)

Latchman D S 2014 *Gene Control* (New York: Garland)

IOP Publishing

Some Critical Questions in Biological Physics
A guided tour around the bugbears
Thomas Andrew Waigh

Chapter 12

The origin of sub-diffusion inside cells—everything's gone fractional

12.1 Introduction

Cells are the fundamental building blocks used to construct living organisms. Inside these cells, a vast range of processes occur that are vital for the physiology of the organisms, such as signalling, analogue calculations, the creation of new molecules and motility. Life is extremely complex, but at the same time it is a robust and reproducible phenomenon. At the molecular level all molecules in solution jiggle around due to thermal energy in a random manner, a process called Brownian motion. Robust living processes emerge from this random motion of the molecules, provided sufficient statistical averaging occurs (from multiple instances of the same molecule or in the long time limit). However, inside cells an unusual type of this random Brownian motion occurs (*sub-diffusive*), due to the extremely crowded environment that exists. Sub-diffusive motion can impact on the majority of the tasks performed inside a cell, since it helps determine reaction rates and the speeds at which cargoes are transported. The quantitative understanding of sub-diffusive motion is far from being complete and its implications for motor proteins, gene regulation, endosomal trafficking, pattern formation etc, are only slowly becoming unravelled.

Quantitative physical models are sometimes developed hand in hand with the mathematical techniques used to describe them and historically this has been a major factor that controls model development. The models created can be either aided or obstructed by the mathematical tools used for their description. Scientists tend to have a working range of skills in mathematics with which they are fully comfortable and a slightly wider range of methods that they are reasonably familiar with, but would need some effort to master them in detail for them to be usefully applied in a real research problem. To move beyond this comfort zone requires much more intellectual effort and entails large amounts of brow furrowing and head scratching. People often need many months of practice, until new mathematics tools

doi:10.1088/978-0-7503-1374-2ch12

© IOP Publishing Ltd 2017

can be competently used and assimilated into their own research fields. Sub-diffusion inside cells appears to require some new mathematical tools to be introduced for a satisfactory description. These tools tend to go beyond the traditional comfort zones inhabited by physicists and engineers, and one of the foremost possibilities is based on fractional differential equations. Such fractional differential equations are extremely mathematically sophisticated.

The title of Metzler and Klafter's classic 2004 review of the subject of sub-diffusion was called *The restaurant at the end of the random walk: recent developments in the description of anomalous transport by fractional dynamics*, following the title of the science fiction book by Douglas Adams, *Restaurant at the End of the Universe*. Famously in Adam's book the answer to life the Universe and everything was known (it was 42), but unfortunately no one knew the specific question to which it was the answer. Metzler and Klafter's answer was known (the analysis of fractional differential equations that describe sub-diffusive motion), but the researchers were still searching for the correct question. The key question may be sub-diffusive motion in a live cell, since it is of critical importance to a wide community of researchers in the life sciences.

Diffusion inside cells will be considered in more detail presently, but first an aside is needed about springs. Physicists love models with springs, since often they can act as a first approximation for force laws between particles and Hooke's simple law can describe them. The forces the springs experience are proportional to their displacements, and life is mathematically very simple. Nevertheless, in reality there is no such thing as a perfectly elastic material; they do not exist. This is as true for bits of rubber band on the author's desk, and the steel shock absorbers in his car, as for the minute bonds that occur inside individual molecules. Even the simplest molecule imaginable is rather unspring-like.

The physics laboratory in Manchester (the Schuster building) is named after Sir Arthur Schuster (1851–1934) who was a renaissance type scientist with an impressively wide range of interests. In addition to investigations on molecular spectroscopy, antimatter (he coined the word) and electrochemistry, Schuster produced one of the first x-ray images to be used in a criminal trial. The pictures were of a bullet hole in someone's head and they were used by the prosecution in a murder trial. Schuster's investigations into early molecular spectroscopy experiments are now a standard part of undergraduates' degree courses during their third year in laboratory modules for present day physicists. The author has frequently demonstrated Schuster's work during term time in Manchester. In the laboratory experiment, light from a car head lamp is passed through a gas of iodine (I_2) molecules and the molecules' absorption as a function of the energy of the light is measured using a spectrometer. The data shows very clearly that the covalent bonds between the iodine atoms act nothing like perfect elastic springs. Such bonds are called *anharmonic*, as opposed to a harmonic bond that does indeed follow Hooke's law. Iodine is one of the simplest molecules that can be considered and it is fairly typical in its behaviour, so it is concluded that all real molecules are anharmonic.

Obviously more sophisticated models are needed to describe anharmonic systems to move beyond primitive Hookean springs. A brute force solution is to simply add on a small correction to the harmonic model and hope this does a better job. This is called a *perturbation theory*. However, often it is worth a complete overhaul of these problems to understand the dynamics of molecules in detail, a back-to-basics revision of the models, since the perturbation theory can be the mathematical equivalent of attempting to force a square peg into a round hole. Perturbation theories can end up with hundreds of algebraic terms. They become extremely unwieldy, and there is often no assurance that calculations based upon them will necessarily converge and give the correct result. Fractal models provide a new tool to describe anharmonicity (fractional harmonic equations) and these models can be expressed as fractional differential equations. Therefore, fractional models are not only required to understand the diffusive jiggling motions of molecules inside cells, they also describe the internal vibrations of the molecules as well, which are also quasi-elastic. These two types of motion of a molecule (translational and internal) are found to be closely related.

Another way to consider Hooke's law is that it is a linear theory, since the force is proportional to the displacement. In contrast, fractional models are inherently non-linear. Additional unique exotic phenomena in the physics of materials that are associated with non-linearity have been found due to anharmonic potentials, such as *solitons*. Examples of solitons include tidal bores found in some rivers (abrupt changes in height that progressively move down stream, e.g. the Severn estuary bore that people have surfed on) and electronic phenomena in DNA. However, in this discussion we will focus upon the slightly more subtle phenomena, which are also associated with anharmonic potentials that determine intracellular transport and are known to impact on almost all cellular processes.

The discussion now returns to the subject of fractals and they will be considered in a little more detail before it returns to intracellular transport inside cells. Fractals have both been successfully used and abused, since their introduction by Benoit Mandlebrot in the 1970s. A slightly facetious example concerns the fractal exponent of Mickey Mouse. Take a picture of Mickey and draw around the perimeter of the mouse. A fractal exponent can then be calculated to describe the wiggliness of Mickey's perimeter. The answer could be 3.9104. Next the question is asked as to what is the fractal exponent of Mickey Mouse when he holds hands with Minnie Mouse and the answer could still be 3.9104. Is Mickey really a fractal? Is the exponent really any use if it cannot differentiate between a single mouse and a mouse combo? Even if the mouse combo exponent was different, what new information is it really telling us? This is an example of the abuse of fractals. An object that clearly is not a fractal (self-similar) on a range of length scales, Mickey Mouse, is being described with a fractal model. The mathematics does not know any better and it will provide an exponent to describe the curve, although this exponent will mean very little. Nevertheless, this should not detract from systems in which fractals really are applicable. Furthermore, the scaling laws that underlie fractal behaviour were around long before fractals were discovered in the 1970s and they are much more general in their scope. Scaling laws are extremely useful mathematical modelling tools.

Fractional differential equations are also closely related to models for the dynamics of particles confined to fractal geometries and such equations have a long history that stretches back to one of the two fathers of calculus, Gottfried Wilhelm Leibniz (1646–1716). Leibniz asked the question as to whether the fractional derivative of a function (f), $\frac{d^{1/2}f}{dt^{1/2}}$, exists. However, except until very recently, fractional derivatives have mostly been overlooked as a viable description of real physical phenomena. The fractional techniques had only been viewed as mathematical curiosities.

Often the mathematics taught to physicists, chemists and engineers is so old it rarely describes 20th century discoveries, let alone more recent 21st century developments; the maths is predominantly restricted to pre-1900 material. In some respects this parsimonious use of mathematics by physicists may have restricted the range of physical phenomena that have been discovered. For example, in abstract terms it can be asked, why do most of the famous laws of nature seem to be describable by linear differential equations? Is it because they are indeed the fundamental descriptions of the world, or does it just indicate our conservative use of mathematics for the construction of physical laws, i.e. it demonstrates a lack of mathematical imagination on the part of the researchers? For example, Maxwell's equations in a vacuum are linear, but the transmission of electromagnetic waves through any real material immediately requires non-linear extensions for a high-resolution description. Why are non-linear differential equations not used more often? It is easy to argue that non-linear equations are neglected because they are hard to handle, not because they are unimportant.

Once a person's eyes are opened to these non-linear possibilities, they start to appear all over the place, e.g. bend an elastic rod (used as a model for a mouse's whiskers in chapter 4, but a plastic ruler on someone's desk will also exhibit the phenomenon) and classically the Euler–Bernoulli equation is needed to describe its elastic response. For anything more than very small deformations, the Euler–Bernoulli equation is a non-linear differential equation. There are many more examples where natural systems described by experimentally verifiable models appear to be non-linear. The fundamental equations in fluid mechanics due to Navier and Stokes are extremely non-linear. So too is oscillatory motion in a dissipative environment, and there are countless other examples of non-linear behaviour that occur in biology, from the electrical activity in the neurons of a brain to the internal biochemical clocks in humans that cause jet lag. Indeed, a wide variety of organisms have circadian (day/night) clocks based on non-linear oscillations and the phenomenon of jet lag has also been demonstrated with bumble bees, i.e. the bees wake up at the wrong time to pollinate flowers if they are moved between different time zones in a plane.

Even more exotic non-linear mathematics can be invoked in the form of fractional differential equations (based on Leibniz's discoveries) to describe physical systems. Such equations are good for coarse graining some non-Markovian processes (statistical processes that have a memory) that represent a wide range of physical phenomena that include sub-diffusive random walks. Coarse graining is just a fancy word for averaging. Some unimportant molecular details are lost during the averaging process. The mathematics of fractional differential equations is however intriguing, since there still is not a universally agreed upon definition in terms of the formal algebraic expressions. It is therefore still an attractive area where new fundamental contributions to the field of mathematics could be made. Unfortunately, the subject is not for the faint hearted. Many of the rote learned rules for the solution of differential equations, such as the chain rule, no longer work with fractional differential equations. Also, the calculations are often much more algebraically involved and they are more easily solved with integral equations, which again can seem foreign and unfriendly to the uninitiated researcher.

So finally, after this long preamble, an extremely fundamental problem for biophysics will be introduced. All particles suspended in water constantly jiggle around due to the thermal energies in the environment and the amount of motion becomes larger as the size of the particles becomes smaller. This was part of Albert Einstein's PhD thesis in 1905 and led to a Nobel Prize for a Frenchman (Perrin) in 1926 when he experimentally demonstrated the results. Nevertheless, Einstein's analysis only works for the diffusion of isolated particles in dilute solutions. Place a diffusing particle in a more congested environment and obviously it moves less far. However, the statistics of the motion are also completely different. In the scientific jargon the motion is called *sub-diffusive*. Sub-diffusive motion is found to commonly occur in living cells. In cells, the statistics of this sub-diffusive motion at small lengths determines the dynamical behaviour at larger lengths scales, but is it the static structures (the labyrinthine architectures in the cell; the ant in a labyrinth model, figure 12.1) that are important, the molecular dynamics (due to the thermally driven motion of the molecules, an ant with an elastic harness, or nanomotors driven by energy stored in chemical fuels, a motorised ant) or a combination of the two? The most successful statistical models to describe this phenomenon can be described by fractional differential equations.

L. *Question*: Determine the origin(s) of sub-diffusive motion in a variety of human cells.

It is also possible to pose a closely related question with regard to sub-diffusive motion for the internal dynamics of proteins (due to anharmonicity). The internal

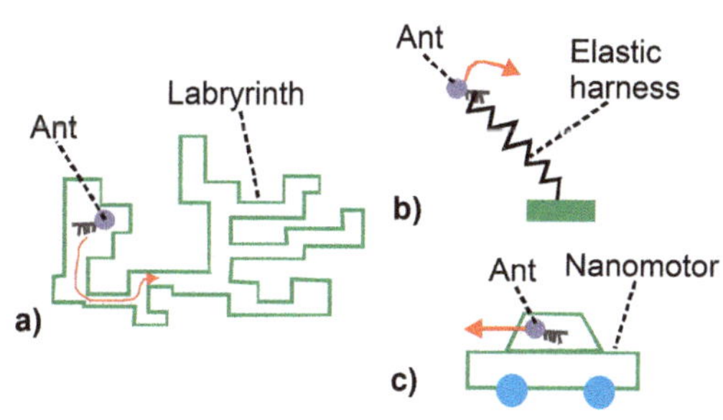

Figure 12.1. Simple models for sub-diffusive motion include (a) an ant diffusing in a labyrinth, (b) an ant diffusing with an elastic harness and (c) a motorised ant.

dynamics that result in their sub-diffusive motion are also significant in all large molecular systems that include carbohydrates, membranes and cellular organelles. Here we focus on globular proteins for concreteness, but there exist many additional questions in the molecular dynamics of sub-diffusion in other biomolecular systems.

L. *Question:* Provide an accurate quantitative model for sub-diffusive motion at the intramolecular level inside proteins.

Elastic vibrations in solids are known to be quantised (i.e. they adopt discrete values for their energy, momentum, number etc that occur in minute chunks) and are described by the theory of quantum mechanics (chapter 14). Such chunks of harmonic vibrations are of central importance in solid state physics where they are used to explain a diverse range of phenomena, such as the switching of transistors, light emission from LEDs and the conductivity of superconductors, and the chunks are called *phonons.* These phonon models for the quantised elastic vibrations due to Hookean excitations inside solids have recently been extended to describe vibrations that occur on fractal networks (e.g. the energy landscapes of single globular proteins can be described by fractals) and are called *fractons.* Such fracton models can provide new insights into the internal dynamics of proteins, and they have been generalized and tested with other materials in different dimensions, e.g. membrane surfaces embedded in three dimensions. Without a good model for anharmonic vibrations inside proteins (or the closely related phenomena of harmonic vibrations on fractal geometries) it is impossible to predict their activity, e.g. how proteins will catalyse a reaction, modify the activity of surrounding molecules or reinforce the mechanics of cellular architectures. Fractons thus provide a new possibility for the development of more realistic models for proteins, although detailed predictions on fractons based on anharmonic bonds relevant to proteins have not yet been made.

Additional important examples of sub-diffusion occur in many other areas of biophysics. For example, new models based on fractional diffusion equations can describe electrical propagation in cardiac muscle, which is an extremely heterogeneous material in terms of its conductivity, and the models are found to be in good agreement with experiment (chapter 18).

In general, a cynic might ask whether the development of the new fractal models that has been advocated are fractional fitting equations or represent new physics. The current consensus is they do contain some new physics, if issues such as those encountered with the model of Mickey Mouse are avoided, i.e. misapplication of the concepts. Diffusion of particles in all environments (both congested and uncongested) clearly is fractal-like in high resolution experiments that cover a wide range of time and length scales (many orders of magnitude are needed to convince non-believers). Fractional differential equations appear to be the foremost technique to describe the statistics of diffusing materials in general and are particularly well adapted to systems where crowding is thought to be important, and when there is an appreciable amount of viscoelasticity in the system (chapter 13). Once fractional differential equations have been properly tamed, they may become tools that are generally used for problems in a wide variety of biological physics research. Indeed, they promise new high-resolution models in the field of cellular biology for the majority of both the molecular motions and the chemical reactions that occur.

12.2 Suggested reading

Introductory

Hernmann R 2011 *Fractional Calculus: An Introduction for Physicists* (Singapore: World Scientific). An expert tour around the physical applications of fractional calculus.

West B J 2016 *Fractional Calculus View of Complexity: Tomorrow's Science* (Boca Raton, FL: CRC). A popular account of the applications of fractional differential equations in a wide range of science problems.

Strogatz S H 2014 *Non-linear Dynamics and Chaos* (Boulder, CO: Westview). This textbook is currently the best introduction to the mathematics associated with non-linear phenomena, and it includes a wide variety of biological examples.

Advanced

Reuveni S, Granek R and Klafter J 2010 Anomalies in the vibrational dynamics of proteins are a consequence of fractal-like structure *PNAS* **107** 13696. Primary research literature in which a case is made for anharmonic fractal vibrations for the internal dynamics of proteins, i.e. fractons.

Hoeflig F and Franosch T 2013 Anomalous transport in the crowded world of the biological cell *Rep. Prog. Phys.* **76 046602**. An extensive review is made of sub-diffusive transport in live cells. The experimental evidence for this behaviour is now overwhelming. It appears to be the rule in the cells of all living organisms, rather than the exception.

Winfree A T 2010 *The Geometry of Biological Time* (Berlin: Springer). Describes new mathematical tools to understand the internal clocks inside living organisms, e.g. jet lag in bumble bees.

Metzler R and Klafter J 2004 The restaurant at the end of the random walk *J. Phys. A: Math. Theor.* **37 R161**. Describes a wide range of models that are used to understand sub-diffusion

Bueno-Orovio A, Kay D, Grau V, Rodriguez B and Burrage K 2014 Fractional diffusion models of cardiac electrical propagation: role of structural heterogeneity in dispersion of repolarization *J. R. Soc. Interface* **11** 1–12

12.3 Technical details

12.3.1 The fractal exponent of Mickey Mouse

Fractals provide a convenient model to describe both structures that are self-similar on a range of length scales and their dynamics on different time scales (Strogatz 2014). No real materials are perfectly fractal, since quantum mechanical chunks inevitably occur on small length scales and no real system is infinitely large, but if the conditions of self-similarity hold over a few orders of magnitude, fractal models are often thought to be useful descriptions of reality. To extend the applicability of fractal models, multi-fractals have been proposed, where the fractal exponent can vary as a function of length scale, but these tend to be more empirical in nature and they are often less convincing as a result (they appear to have been cobbled together and only describe specific data sets).

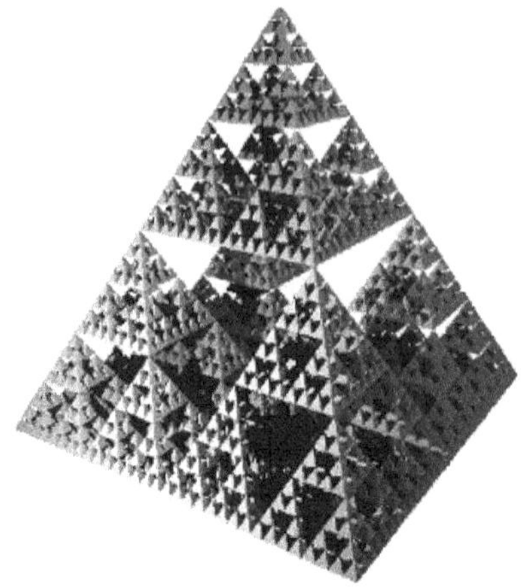

Figure 12.2. A well-defined mathematical model is used to describe the Sierpinski gasket, which is a three-dimensional fractal. Mathematicians have derived equations to describe Brownian motion on such structures and they can be used to describe the motion in cells (graphics courtesy of Paul Bourke, http://paulbourke.net).

An example of a fractal in three dimensions is shown in figure 12.2, called a Sierpinski gasket. This might seem a fairly artificial construct for a biological system, but such architectures can provide valuable reference models, and exact analytic results can sometimes be derived for them.

One approach to motivate the sub-diffusive behaviour observed for particle transport in high concentration systems is to consider the motion of an ant inside a fractal maze (figure 12.1). The discoverer of fractals, Benoit Mandelbrot, published extensively on this subject and the models have been adopted by physicists to describe some naturally occurring phenomena. Whether Brownian motion on a fractal occurs in biological cells is still open for debate. The contents of cells have high concentrations and motile particles will need to negotiate their way around surrounding particles if they want to fulfil their functions, so some kind of labyrinth exists. However, it is not clear why these excluded volume effects should be self-similar over a broad range of lengths scales, i.e. perfectly fractal.

12.3.2 Definition of sub-diffusive motion

The basic definition of sub-diffusive motion for particles can be used to determine whether it occurs in experiments and it is very straightforward. Mean square displacements ($\langle \Delta r^2 \rangle$) are used to categorise the motion and they are observed to follow a simple power law behaviour, i.e.

$$\langle \Delta r^2 \rangle \sim \tau^{\alpha} \tag{12.1}$$

where $\langle \Delta r^2 \rangle$ may or may not be time averaged depending on the context, τ is the time interval between displacements and α is a positive real number. Motion is defined as sub-diffusive when $\alpha < 1$, diffusive when $\alpha = 1$ and super-diffusive when $\alpha > 1$ (figure 12.3) (Waigh 2014, Kenwright *et al* 2012). Additional energy, beyond the standard amount due to the thermal environment (kT), is required to drive the

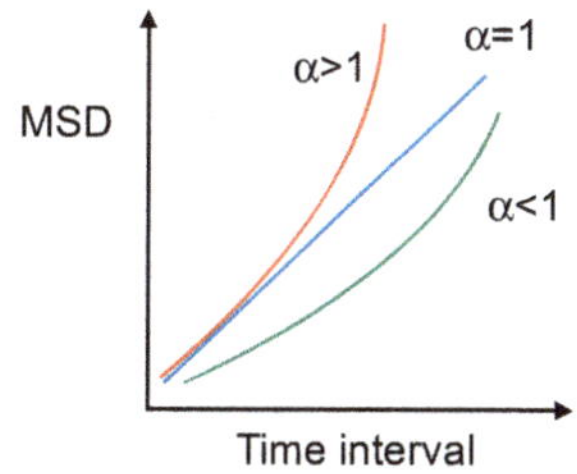

Figure 12.3. Mean square displacement as a function of time interval for particles obeying, diffusive ($\alpha = 1$), sub-diffusive ($\alpha < 1$) and super-diffusive ($\alpha > 1$) motion (12.1).

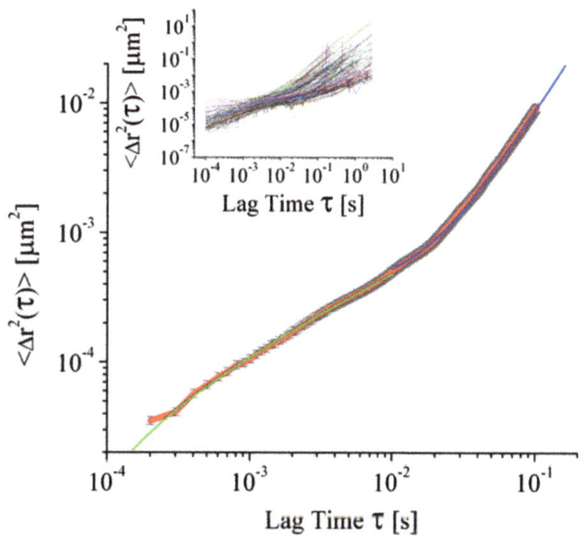

Figure 12.4. Mean square displacements ($\langle \Delta r^2(\tau) \rangle$) as a function of time interval (τ) for endosomal particles inside human lung cells. The inset shows MSDs for individual intracellular endosomes as a function of time interval. Reality is often a lot more complicated than implied by theory and two different power laws 12.1 are observed at different time scales. The MSDs are sub-diffusive at short times ($\alpha < 1$, $\tau < 10^{-2}$ s) and super-diffusive ($\alpha > 1$, $\tau > 10^{-2}$ s) at long times scales due to the activity of motor proteins. Reprinted from Harrison *et al* 2013 with permission from IOP Publishing.

motion during super-diffusive transport, e.g. fuel is needed to drive nanomotors or an external convective current is needed.

Sub-diffusive motion is mostly ignored in current models for intracellular motility, due to the relative complexity of the calculations. Data from our own group (figure 12.4) clearly demonstrate sub-diffusive motion for individual intracellular endosomes at short times ($<10^{-2}$ s). It appears to be a robust phenomenon and it is not simply due to averaging of Brownian trajectories over a heterogeneous population of motile particles.

Sub-diffusive transport is often accompanied by reactions inside live cells and these reactions therefore require sub-diffusive reaction-diffusion equations for an accurate description. Good mathematical predictions exist for the motile behaviour that is expected, but the current challenge is to determine whether they provide useful robust predictions for experiments (Mendez *et al* 2012).

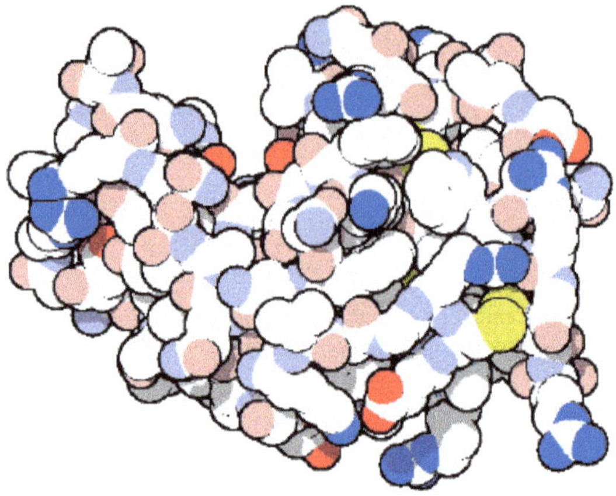

Figure 12.5. Diagram of the structure of a globular protein, lysozyme, from the protein database, based on data calculated using x-ray crystallography.

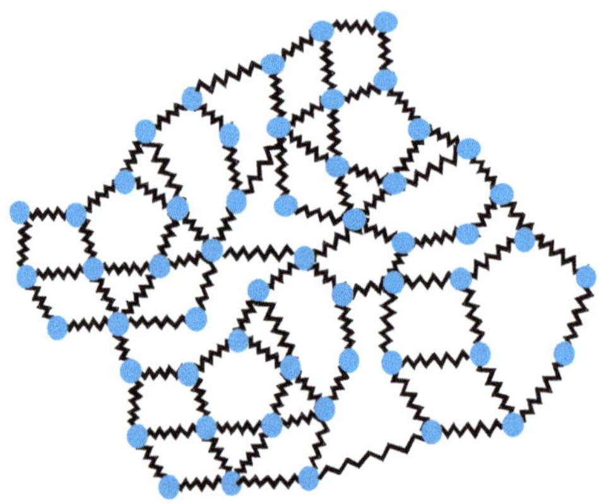

Figure 12.6. Bead spring network model for proteins are fairly well developed in the case of Hookean springs. There are also useful tools available, such as principal component analysis, to classify the spectrum of vibrations of the springs.

12.3.3 Protein springs

Globular proteins have complex geometries that provide a challenge for would-be modellers (figure 12.5). A schematic diagram of a mechanical model of a globular protein is shown in figure 12.6. To build models for the globule's dynamics this standard type of toy model replaces each amino acid in the chain by a sphere and then models their interactions with surrounding amino acids using harmonic springs. Such models are particularly popular as tools to describe spectroscopy experiments, e.g. the results of inelastic neutron scattering experiments. To date some progress has been made considering the impact of fractal energy landscapes for the proteins on their dynamics (Neusius *et al* 2008, Reuveni *et al* 2010). However, advances in anharmonic potentials would clearly improve the accuracy of such models, since all real potentials are in general anharmonic.

References

Harrison A W, Kenwright D A, Waigh T A, Woodman P G and Allan V J 2013 Modes of correlated angular motion in live cells across three distinct time scales *Phys. Biol.* **10** 36002

Kenwright D A, Harrison A W, Waigh T A, Woodman P G and Allan V J 2012 First passage probability analysis of active transport in live cells *Phys. Rev.* E **86** 31910

Mendez V, Fedotov S and Horstemhe W 2012 *Reaction-transport Systems* (Berlin: Springer)

Neusius T, Daidone I, Sokolov I M and Smith J C 2008 Subdiffusion in peptides originates from the fractal-like structure of configuration space *Phys. Rev. Lett.* **100** 188103

Reuveni S, Granek R and Klafter J 2010 Anomalies in the vibrational dynamics of proteins are a consequence of fractal-like structure *PNAS* **107** 13696

Strogatz S H 2014 *Non-linear Dynamics and Chaos* (Boulder, CO: Westview Press)

Waigh T A 2014 *The physics of living processes: a mesoscopic approach* (Chichester: Wiley)

IOP Publishing

Chapter 13

Microrheology—the unexplored continents

13.1 Introduction

A walk across the surface of a swimming pool filled with custard is indeed possible and it is particularly popular on South American variety television shows, if movies on YouTube are to be believed. Surprisingly, the physics behind this wacky phenomenon is still not completely understood to a quantitative level of detail. It is known that suspensions of custard particles in water become elastic, if deformed quickly when a person walks at a brisk pace, due to jamming of the particles, and the custard can then sustain the person's weight. However, during slower deformations, when people stop walking, the jams become unblocked, the suspension becomes fluid-like again and people will sink into the custard. Technically this viscoelastic phenomenon depends on time and is called *shear thickening*; as the material is deformed with time (sheared), it becomes thicker and more elastic. The custard experiment is a first example of a viscoelastic phenomenon, but there are many more.

All condensed materials, with the possible exception of liquid helium at very low temperatures, are to some degree viscoelastic, so there is a huge range of possibilities. There are even some theories that describe viscoelastic phenomena in helium superfluids, but it is still a strongly contested area. The experiments present many challenges due to the extremely low temperatures required, and the quantum physics can be very subtle.

Even water molecules at room temperature bounce around with elastic deformations at picosecond time scales due to the finite time for bond rearrangements (classically this is surprising, because water was often considered as a prototypical fluid), whereas mountains (such crystalline materials are standard examples of elastic solids) can flow over a duration of thousands of years. Water droplets can also be made to bounce from super-hydrophobic coatings at millisecond time scales, but this is a separate phenomenon, driven by the surface interactions and not the drop elasticity *per se*. Examples of hydrophobic coatings include polishes used for cars and waxy crystalloids on the surfaces of lotus leaves. Defect structures in

doi:10.1088/978-0-7503-1374-2ch13 © IOP Publishing Ltd 2017

crystalline materials are often relatively mobile, which allows them to flow very slowly when a stress is applied (to understand the occurrence of slowly flowing mountains) and explains the related phenomenon of ductility in metals, e.g. why sword smiths repeatedly strike their weapons on anvils to harden them, which reduces the mobility of the defects, makes it harder for the metal to flow, and consequently increases the weapons' strengths.

Silly putty and tomato ketchup are other standard examples of viscoelastic specimens and they can be easily explored over the range of time scales commonly probed by human senses. Throw silly putty at a floor (which is assumed smooth and rigid) and it bounces like an elastic ball, whereas pull it slowly in the hand and it flows like a fluid. This weird behaviour is because the molecular relaxation times for the unusually large molecules in silly putty are very slow and happen at time scales on the order of seconds. Deform the silly putty molecules faster than their relaxation time and the molecules do not have time to flow and they will bounce elastically. In contrast, tomato ketchup often displays an obstinate desire to remain in its bottle if someone thumps on the bottom of the bottle. Continued thumping eventually causes a sudden fluidisation of the ketchup and the resultant sudden increase in flow is hard to control. This is called *thixotropy*; a fluidisation phenomenon that depends on time. More specifically, thixotropy demonstrates hysteresis in its behaviour, i.e. the fluidisation phenomenon depends on the past stress history of the sample. Chips can easily end up drowned in a sea of ketchup, unless this process of thixotropy is carefully controlled. The molecular origin of this phenomenon is fairly subtle, associated with the disruption of inter-particle forces, and it is still an active area of study in the field of colloidal hydrodynamics.

The research field that describes the study of viscoelasticity of materials is called *rheology*; from the Greek word for the study of flow. Rheology experiments often involve the use of a rheometer; a device that is the scientific equivalent of a food blender. For example, a slab of a material could be placed between two plates in a rheometer, figure 13.1. The upper plate could be made to oscillate, applying a stress to the material with a well-defined frequency. The response of the material to the stress (e.g. its angular displacement) can be used to quantify its viscoelasticity as a function of frequency. With silly putty at high frequencies a relatively stiff response

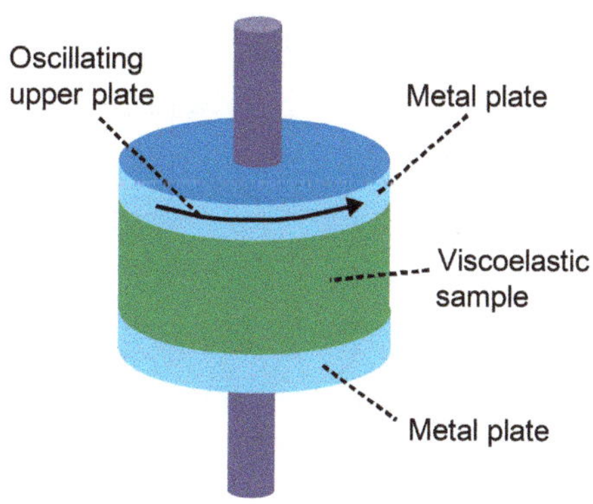

Figure 13.1. A simple plate/plate rheometer for soft condensed matter can be used to study the viscoelasticity of a material, i.e. its rheology.

occurs (a high shear modulus), whereas at low frequencies a relatively soft response occurs (a low shear modulus).

The author did not have much money in the laboratory in which he was given his first permanent academic position at the University of Leeds. Most of the experiments performed therefore needed to be as cheap as possible. This financial constraint led to some microrheology experiments that involved an optical microscope, which was loaned from a colleague, small polystyrene particles suspended in water, a video camera and a video recorder (it was a DVD player, now an outmoded technology). To modern eyes this may seem to be a very low technology approach, but similar experiments led to a Nobel Prize for a Frenchman, Jean Baptiste Perrin in 1926, so this type of experiment has proven itself to have a good pedigree in terms of a historical context. Perrin investigated minute rubber spheres that were suspended in simple Newtonian fluids and manually tracked their motion using a microscope. He used his results to test some of Einstein's predictions on Brownian motion, which in turn provided excellent new evidence for the molecular nature of matter (still open for debate in the 1920s). Perrin was also able to nail down an accurate value for Avogadro's number from his measurements, which is one of nature's fundamental constants. The current challenge is to extend this type of tracking measurement to viscoelastic specimens; a field known as *microrheology*.

Rheology is the study of the flow behaviour of materials (that are normally viscoelastic) and the field of microrheology considers the flow properties of very small (microliter) quantities of sample. To date, microrheology methods have only investigated a few micro-islands among the vast continents of possible experiments into viscoelasticity, and there are still an enormous number of systems left to be explored. The process of exploration is in part restricted by the development of the necessary instrumentation, since up to now only a relatively small range of moduli and time scales can be probed. Current microrheology techniques are mostly limited to very soft viscoelastic fluids and gels due to their limited sensitivity to materials with high shear moduli (most solids), whereas their time sensitivity is very competitive (they can explore processes from thousands of seconds to microseconds). To be more specific, the simplest microrheology experiment uses an optical microscopy with a viscoelastic fluid, figure 13.2. Polystyrene spheres

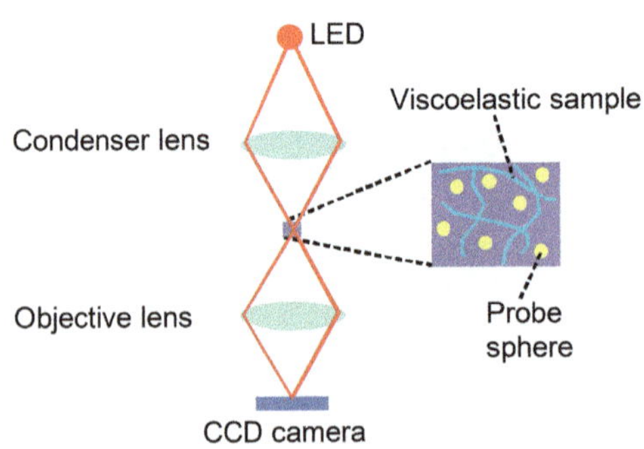

Figure 13.2. Simple experimental arrangement for a particle tracking microrheology experiment inside a viscoelastic fluid (a solution of polymers in this case).

(~1 μm) are embedded in the viscoelastic sample and their motions are tracked in a microscope. A clever piece of statistical physics means that we know the range of stress applied by background thermal energy to the spheres as a function of time, since they follow a white noise profile. Knowing these applied stresses and measuring the resultant responses of the spheres (their displacements) means that we can calculate the viscoelasticity of the material in which the spheres are embedded using a sophisticated theory (the Generalized Stokes–Einstein equation).

Even with these restrictions on the microrheology experiments possible presented by the range of moduli and time scales, the fact that the majority of materials are to some degree viscoelastic, means there is still a huge choice for possible systems to study. Therefore, only a few important challenges for biological physics in the field of microrheology will be considered in this chapter; a much more limited subset of all the possible types of materials that could be explored.

First of all consider the viscosity of *polyelectrolytes*. Polyelectrolytes are a very common class of molecule in biology and there are a series of important questions associated with their viscoelasticity, e.g. the dynamics of hyaluronan in the knees during walking, amyloid aggregates in brain disease, mucin in the stomach that affects ulcers or pectins placed in yoghurts to help improve their shelf-life (chapter 17). Hyaluronan injections are often a first treatment for osteoarthritis, since the patients can then avoid the need for a full joint replacement operation for a few more years, which are very expensive, painful and invasive procedures. The chemistry of hyaluronan is fairly straightforward and well understood, so the optimisation of its viscoelasticity is a clear pragmatic goal for bioengineers that can be achieved with inputs from rheology experiments.

There already exists a reasonably serviceable scaling theory of polyelectrolytes in the literature that describes their viscoelasticity. Scaling theories can describe how the physics depends on the key parameters in a problem, but they tend to be agnostic with respect to fully quantitative predictions; they are only broad brush stroke models. Nevertheless, the polyelectrolyte scaling theory fails fairly dramatically for short polyelectrolyte chains and never predicts a sufficient amount of quantitative detail that people can say that they are really completely convinced by it with long polyelectrolyte chains. The theory is clearly partially wrong and conceivably it could be totally wrong.

L. *Question*: Create a good quantitative theory for polyelectrolyte viscoelasticity and test it with some high-resolution microrheology experiments.

Another field of interest that has experienced some rapid developments is that of the viscosity of motor protein solutions. Mix some motor proteins (e.g. myosin molecules from rabbit muscle; molecules purified from rabbit meat purchased at a local butcher will suffice) with solutions of the complementary molecular rail road tracks (e.g. linear actin chains that cross-link them into networks) and weird things are observed in the solutions. Given an appropriate source of fuel for the motors (e.g. ATP), the myosin molecules begin to inch worm their way along the actin tracks. This behaviour is well established in the field, but strange swirling patterns that are self-driven within the fluid can also be observed under an optical microscope and the patterns resemble those seen inside living cells when they divide. The motor protein solutions are called *active liquid crystals* and as such represent a new class of condensed matter. These rabbit systems are intriguing and may help drive some fundamental research, but appear a little erudite.

A clearer application for motor protein viscoelasticity can be seen in the behaviour of cardiac cells. *Rigor mortis*, which translates from Latin as *stiffness of death* (with all its accompanying slang words and phrases, such as America's *stiff* and Shakespeare's *dead as a door nail*), is due to the cessation of motor protein activity in muscles upon death. Fixed cross-links between the molecules in dead muscle cause a large increase in both a body's elasticity and viscosity.

L. *Question:* Create a quantitative model for motor protein viscoelasticity in living muscle and in dead muscle (in *rigor mortis*), based on the activity of the motor proteins using high resolution microrheology experiments.

Such a model would be extremely useful for the development of an understanding of cardiac disease, e.g. how viscoelasticity relates to the efficiency with which a heart pumps when part of the heart tissue dies (an infarction caused by a heart attack), that can in turn be affected by numerous other medical conditions, such as diabetes.

Another extremely fundamental medical issue that needs attention in microrheology experiments is the viscoelasticities of the entire range of genes that express mucin in humans (chapter 6). There are 21 mucin genes that are currently known to occur in humans and each one codes for a material with different viscoelastic properties. The expectation is that mucin viscoelasticities are finely tuned to their functions, e.g. the mucins that protect the front of the eyeball from shearing motions when the eyelid closes require a different viscoelasticity to that needed to protect the wall of the stomach against ulceration (due to autodigestion). The functional mechanisms of mucins are critical in a wide range of medical conditions, e.g. lung disease, tooth decay and gastrointestinal disease.

L. *Question:* Measure the viscoelasticity of all 21 human mucin genes as a function of the concentration of the solutions, i.e. create dynamic phase diagrams for each type of mucin with a high level of resolution for their time responses.

One final medically relevant application of microrheology that is worthy of consideration is in microbial biofilms. Such biofilms protect the bugs from antibiotics and are a major contributor to the problem of antibiotic resistance. The Armageddon scenario is that a wide range of bacteria will soon become antibiotic resistant and medical science will be put back by 100 years (chapter 10). Common everyday infections will become untreatable and lead to a large number of causalities. Microrheology provides a neat method to monitor the integrity of biofilms through the use of the film's viscoelasticity and it can thus gauge the efficacy of methods to disrupt them. Such microrheology measurements could be used by pharmaceutical companies as a high throughput method to test antimicrobial drugs. These rheological studies have already provided a number of fundamental insights into biofilm behaviour, e.g. why there must be a yield stress for the biofilms (the biofilms act like elastic solids that require a minimum force to be applied before they begin to flow), otherwise the films would be easily displaced by an imposed stress during cleaning strategies, e.g. dental plaque would be quickly washed off the teeth during a rinse, without the need for prolonged brushing (as per every dentist's advice).

L. *Question:* Measure the viscoelasticity of a range of microbial biofilms (both bacterial and fungal) and make a model that relates their viscoelasticity to the microstructure of the biofilm.

This specific task has a huge scope due to the immense number of varieties of bacteria and fungi on planet Earth ($\sim 10^{19}$ bacterial species, $\sim 10^{6}$ fungal species) and the fact that the majority of them inhabit biofilms. To make it a pragmatic task we thus need to limit ourselves to clinically relevant varieties.

Microrheology studies have also started making forays into zoology. This research has ranged from clearly useful studies, such as the viscoelasticity of the glue that blue bottle flies use to stick themselves onto surfaces (relevant to the development of new pesticides) and the viscoelasticity of the aqueous humour in bovine eyeballs (relevant to ophthalmology), to slightly more weird and wonderful experiments that include the viscoelasticity of mucilage used to trap insects in pitcher plants and the variation of the viscoelasticity inside jellyfish as they mature. Older jellyfish are found to be tougher than younger specimens; it should have been a shoo-in for an Ignoble prize a few years ago. So little has yet been done in the field of zoological microrheology that it is easy to come up with additional unexplored systems to while away one's time in a laboratory, such as the viscoelasticity of frog spawn, hagfish slime, fungal biofilms… the potential list is huge. In addition to their fundamental biological interest, such studies could provide biomimetic inspiration for the creation of new synthetic materials that have viscoelasticities that are similar to their zoological analogues.

The field of microrheology is yet to properly enter into the popular scientific imagination. Francis Crick is perhaps the most famous 20th century scientist in this area (the man who went on to discover the structure of DNA with James D Watson). However, he abandoned his work in the related subject of magnetic tweezers in 1950 to do a PhD on the structure of proteins (during which he discovered the structure of DNA), never to return to the subject, which has probably done it no favours. Crick manipulated magnetic particles inside cells (the technique has subsequently become much easier with the development of computers, image analysis techniques and better defined chemistry for the magnetic beads) and observed their intracellular microrheology. This field of intracellular microrheology has recently experienced a renaissance providing important insights into metabolic events and their diseased states, e.g. changes to the viscoelasticity are found in cancerous cells.

There are still some technical issues, which need to be solved to allow microrheology measurements to be possible with all materials. Nevertheless, the dream of mechanical spectroscopy experiments for all soft materials that can be performed over a broad range of frequencies is very close to a reality and it may help to revolutionize a wide selection of biomaterials research. Hopefully many of the unexplored continents in microrheology research will therefore soon be visited and eventually conquered. The current process of exploration has hardly scratched the surface of the vast range of possibilities and it will be helped by both improved instrumentation and theory.

13.2 Suggested reading

Introductory

Piazza R 2011 *Soft Matter: The Stuff that Dreams are made of* (Berlin: Springer). A well written popular account of soft matter physics.

Davies S 2013 *The Drugs don't Work: A Global Threat* (London: Penguin). A popular account of the global health problem presented by antibiotic resistant microorganisms.

Advanced

Waigh T A 2005 Microrheology of complex fluids *Rep. Prog. Phys* **68** 685–742. A review article that describes the development of microrheology techniques and the physical phenomena associated with viscoelasticity as a function of length scale.

Waigh T A 2016 Advances in the microrheology of complex fluids *Rep. Prog. Phys* **79** 74601. A second review of advances in microrheology, achieved during the ten years since the previous review article was written.

Sanchez T, Chen D T N, DeCamp S J, Heymann M and Dogic Z 2012 Spontaneous motion in hierarchically assembled active matter *Nature* **491** 431–4. Primary research literature that describes the self-driven swirling patterns that are created by mixtures of motor proteins from a rabbit.

Dobrynin A V, Colby R H and Rubinstein M 1995 Scaling theory of polyelectrolyte solutions *Macromol* **28** 1859–71. Primary research literature that presents a state of the art theory for polyelectrolyte viscoelasticity that unfortunately fails for short chains.

Izzo D, Cloitre M and Leibler L 2014 The viscosity of short polyelectrolyte solutions *Soft Matter* **10** 1714. A new quantitative theory for short polyelectrolytes that highlights some of the challenges encountered.

Peisker H, Heepe L, Kovalev A E and Gorb S N 2014 Comparative study of the fluid viscosity in tarsal hairy attachment systems of flies and beetles *J. R. Soc. Interface* **11** 752

Gambini C, Abou B, Ponton A and Cornelissen A J M 2012 Micro- and macrorheology of jellyfish extracellular matrix *Biophys. J.* **102** 1–9

Erni P, Varagnat M and McKinley G H 2008 Little shop of horrors: rheology of the mucilage of *Prosera sp.*, a carnivorous plant *AIP Conf. Proc* **1027** 579. A range of exotic applications of microrheology in zoology is presented in these three articles that includes blue bottles, jellyfish and carnivorous plants.

Rogers S S, Van de Walle C and Waigh T A 2008 Life cycle of bacterial thin films *Langmuir* **24** 13549–55. A primary research article which describes one of the first applications of microrheology to measure the viscoelasticity of bacterial biofilms.

13.3 Technical details

13.3.1 Viscoelasticity of silly putty

To gain an intuitive understanding of the unusual behaviour of viscoelastic materials silly putty is often used. Silly putty is sold as a child's toy and it is predominantly based on the polymer, polysiloxane. It is possible to study the viscoelasticity of this material in detail. An example is shown in figure 13.3. Very simple measurements of the force as a function of extension show a non-linear behaviour (shear thickening) and the force experienced depends sensitively on the shear rate imposed.

13.3.2 Swirling motor proteins

Concentrated mixtures of motor proteins can have unusual viscoelastic properties. Figure 13.4 shows optical microscopy images of actin solutions mixed with myosins (nanomotors). The static images shown in the figure resemble synthetic polymeric liquid crystals with well-defined orientations and defect structures. However, these structures are dynamic, due to an interplay between orientation, the defects and the activity of the motors. The actin/myosin mixtures are classified as a new variety of soft-condensed matter, *active liquid crystals*. They resemble some of the intracellular structures observed inside living cells.

13.3.3 Mucin viscoelasticity

A key question for the viscoelasticity of biological materials is how to connect the molecular structure of mucins to the dynamics of their solutions. Figure 13.5 shows results from our group on the microrheology of purified mucin molecules from the stomachs of pigs (specifically, the mucins are defined by their genetics as MUC5AC). At low concentrations (1–10 mg mL^{-1}) the solutions show a $\sim c^{0.5}$ scaling of the viscosity as a function of mucin concentration which is typical of overlapping

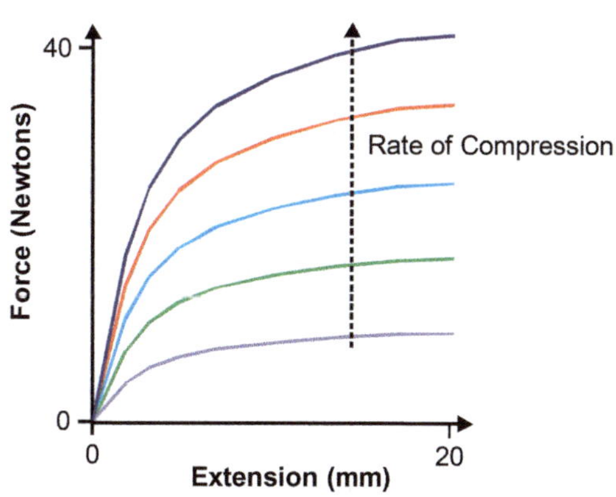

Figure 13.3. The compressive force (F) applied to a cylinder of silly putty as a function of displacement (x). The force is applied at different rates. The faster the force is applied, the larger the force needed to compress the sample, i.e. a time dependent shear thickening viscoelastic phenomenon occurs (Cross 2012).

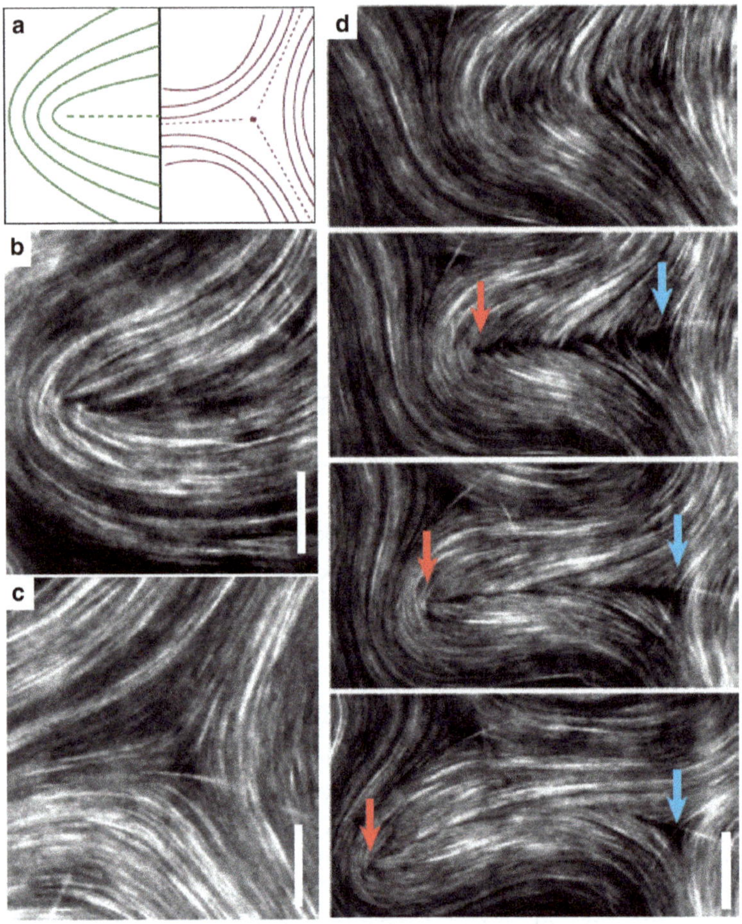

Figure 13.4. Actin proteins mixed with myosin motors (with a small fraction of cross-linkers) give rise to an active liquid crystalline phase. The red and blue arrows show the motion of defects in the orientation of the liquid crystal (Sanchez *et al* 2012). Reprinted by permission from Macmillan Publishers Ltd

solutions of flexible polyelectrolytes (called the *Fuoss law*). However, at around 20 mg mL^{-1} the viscosity has a sudden increase in the scaling of the mucin concentration $\sim c^{3.92}$, which we believe is due to *reptation* (snake-like motion of the polymers). A detailed quantitative explanation of the dynamics of comb polyelectrolytes is lacking, but some scaling laws have been proposed by our group (section 13.3.4).

13.3.4 Scaling model for polyelectrolyte comb viscoelasticity

Scaling predictions from our group for the viscosity (η), relaxation time (τ), shear modulus (G) and the mesh size (ξ) of both comb polyelectrolytes and neutral comb polymers are shown in figure 13.6. A rich variety of dynamic states are possible for comb polymers. Only a small fraction of the complete range of possibilities is shown on figure 13.6, which only compares results for neutral and charged flexible combs

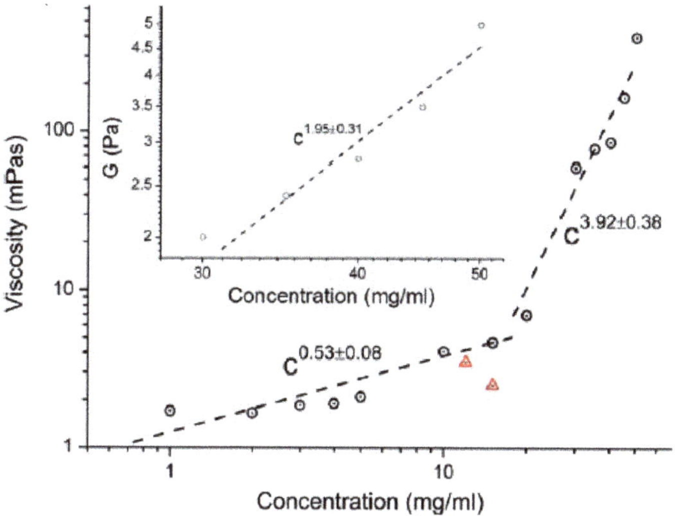

Figure 13.5. Viscosity as a function of mucin concentration for MUC5AC (porcine stomach) mucins. Experiments were performed using particle tracking microrheology (Georgiades *et al* 2014). The inset shows the elastic shear modulus (G) as a function of mucin concentration.

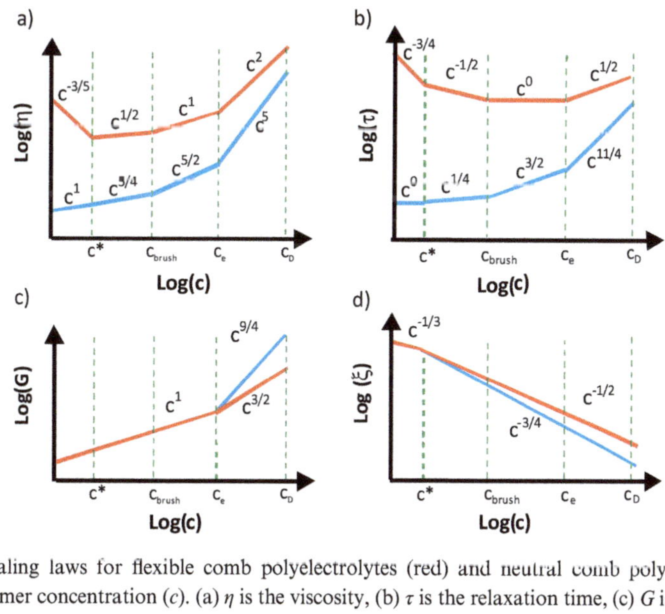

Figure 13.6. Scaling laws for flexible comb polyelectrolytes (red) and neutral comb polymers (blue) as a function of polymer concentration (c). (a) η is the viscosity, (b) τ is the relaxation time, (c) G is the elastic shear modulus and (d) ξ is the mesh size. c^* is the overlap concentration, c_{brush} is the overlap concentration, c_e is the entanglement concentration and c_D is the concentrated solution concentration (Papagiannopoulos *et al* 2008).

with relatively short side-chains. Results from experiments for the viscoelasticity of comb polyelectrolytes, such as aggrecan and mucin, qualitatively fit into this theoretical framework, but an accurate quantitative description (beyond scaling) is not available (Waigh and Papagiannopoulos 2010).

References

Cross R 2012 Elastic and viscous properties of silly putty *Am. J. Phys.* **80** 870

Georgiades P, Pudney P D A, Thornton D J and Waigh T A 2014 Particle tracking microrheology of purified gastrointestinal mucins *Biopolymers* **101** 366–77

Papagiannopoulos A, Waigh T A and Hardingham T E 2008 The viscoelasticity of self-assembled proteoglycan combs *Faraday Discuss.* **139** 337–57

Sanchez T, Chen D T N, DeCamp S J, Heymann M and Dogic Z 2012 Spontaneous motion in hierarchically assembled active matter *Nature* **491** 431–4

Waigh T A and Papagiannopoulos A 2010 Biological and biomimetic comb polyelectrolytes *Polymers* **2** 57–70

IOP Publishing

Some Critical Questions in Biological Physics
A guided tour around the bugbears
Thomas Andrew Waigh

Chapter 14

Quantum phenomena in biology—the role of chunks

14.1 Introduction

At small length scales the world is divided up into minute chunks. Like it or loath it, and regardless of the counter-intuitive nature of some of the predictions of the quantum mechanical models that are used to describe these chunks, this is how reality appears to work and any detailed description of the real world needs to agree with these notions, i.e. new theories must be made that are compatible with quantum mechanics (the mechanics of chunks). Countless experiments have tested the often crazy predictions of quantum mechanics and it is still the most successful physical theory ever created, bar none. Weird predictions of the theory of quantum mechanics include the ability to teleport particles, for a single particle to be in two or more places at the same time, the existence of spin (that unlike a spinning ball, has no spatial extent), that all measurements necessitate significant changes to particle properties… the list goes on!

Quantum calculations that describe the structures of simple atoms and very simple molecules, such as atomic hydrogen (H), molecular hydrogen (H_2), helium (He) and lithium ions (Li^+), are extremely satisfactory, with predictions for their physical properties that agree with experiments to better than ten decimal places of precision. These are thus some of the best physical models that have ever been created in science, and as such are exceptionally successful. Nevertheless, for more complicated atoms (with larger numbers of electrons) and molecules (the majority of matter on planet Earth and predominantly the materials that are of interest in biology), the situation becomes much more dicey and that unfortunately does not just mean statistical theories need to be used. In general, no satisfactory quantum mechanical model has been created for most types of molecules in existence, be they biological or synthetic, and relatively limited from first principles (*ab initio*) predictions can only be made in a very restricted number of situations. Thus, in

doi:10.1088/978-0-7503-1374-2ch14

© IOP Publishing Ltd 2017

general only rudimentary models are available for the quantum chemistry of biological molecules, which is highlighted by current state-of-the-art theories for water.

Humans are 70% composed from water, which helps control the shapes and the dynamic behaviours adopted by all of their constituent molecules. Water is in principle a relatively simple small molecule (H_2O, two hydrogen atoms stuck to a single oxygen atom). Therefore, much effort has been expended in the creation of a quantum mechanical model to explain water's behaviour. Although water is a very familiar substance it has a number of unusual properties, which are fairly weird when given closer inspection, e.g. it becomes less dense when it freezes (opposite to most other materials) and ice thus floats on liquid water. Other unusual properties of water include its extremely high heat capacity, dielectric permittivity and surface tension. The hydrogen nucleus has the highest charge density of any element, since it only consists of a single positively charged proton. This means that bonds of hydrogen to other atoms can become extremely polarised and these polarised hydrogen-bonded atoms thus attract atoms in neighbouring molecules that are oppositely charged by polarisation (hydrogen bonds within a single molecule are also possible). The resultant force is given a specific name, *hydrogen bonding*, due to the unusually high strength of the interaction and it qualitatively accounts for some of water's unusual properties. Unfortunately, a self-consistent quantum mechanical model for all of water's unusual properties does not yet exist. A pessimistic evaluation is that it is an intractable many-body problem and no solution is in fact possible. However, even partial solutions, which only quantitatively describe an aspect of water's behaviour, would be very useful. Here the approach of *phenomenological boot strapping* is advocated, i.e. the use of fits to experimental data to bridge the logical gaps in intractable calculations due to the many-body problem (chapter 17).

Recent advances that hold lots of promise for breakthroughs with hydrogen bonds are provided by infrared spectroscopy experiments that use lasers that are pulsed at femtosecond time scales. Electronic fluctuations in hydrogen bonds occur extremely quickly, but the ultrafast time scales that can be probed by femtosecond (10^{-15} s) lasers can capture many useful details and the energies of infrared photons are well matched to the energies of hydrogen bonds, which maximises the information available. A battery of other complementary experimental methods is also now available, such as femtosecond electron scattering, inelastic neutron scattering and other fast spectroscopic techniques (microwave, terahertz, Raman, dielectric etc), which can provide data with extremely high time resolutions to test models for the dynamics of hydrogen bonds. Armed with some of this newly available high-resolution data the suggestion is:

L. *Question*: Create an accurate quantum mechanical model for a single hydrogen bond between two atoms.

One technical requirement on the road to a better model for hydrogen bonds is the development of superior theoretical techniques to describe anharmonic interactions (chapter 12), i.e. the creation of accurate numerical methods to describe the energy of the bonds. Hydrogen bonds are anharmonic and improved tools to describe anharmonicity could improve the accuracy of simulations.

L. *Question*: Create an accurate model to describe the quantum mechanics of anharmonic potentials.

Once a good model exists for a single hydrogen bond, a natural extension is to consider the cooperative behaviour of many hydrogen bonds that strongly interact with one another. However, this step can be a bit of a nightmare with molecules that experience strong hydrogen bonding interactions, such as liquid water. The hydrogen bonds can bifurcate (an atom can experience two bonds at the same time, which involves the interaction of three or more atoms, whereas the majority of simulation strategies only consider pairwise interactions, i.e. they can only handle the interaction between two atoms at the same time), and both the protons and electrons involved can exhibit extreme quantum phenomena, such as *entanglement* and *tunnelling*. Higher order interactions, such as trifurcations, can also occur with hydrogen bonds. *Entanglement* is such a weird quantum phenomenon that it almost defies non-mathematical description. The physical properties of particles can become intimately connected, such as their spin, so the measurement of one particle's spin instantaneously affects that of another, e.g. quantum teleportation is possible.

L. *Question*: Create an accurate quantum mechanical model for the collective properties of liquid water.

Armed with a reasonable quantum mechanical model for water, the next extension is to consider the interaction of water molecules with very simple ions, such as potassium (K^+), sodium (Na^+) and chlorine (Cl^-). Such ions occur in all living organisms and play a crucial role in many biological phenomena, e.g. inside the ion channels that transmit spiking voltages along neurons in the brain. Inconveniently, such ions interact very strongly with water and the ions therefore carry around with them a shell of partially structured water molecules. The structural effects of these water shells are felt in many biologically relevant phenomena, such as the solubility of proteins in different types of salt solution, the surface tension of salt solutions and the amount of protein denaturation that occurs in salt solutions. They are collectively called the *Hoffmeister series* after their discoverer, since the different structural effects of the ions can be arranged in an ordered list in terms of their efficiency in structuring water (a series) e.g. calcium ions cause proteins to have a lower solubility than lithium ions, which in turn have a lower solubility than potassium ions. A good quantitative model for the Hoffmeister series still does not exist.

L. *Question*: Create an accurate quantum mechanical model for the Hoffmeister series for the structuration of water molecules around simple ions.

Other more exotic quantum mechanical phenomena lurk at small length scales. Except for hydrogen (which only has one electron), entangled electrons exist in every other type of atom. The process of *quantum coherence* (which is closely related to quantum entanglement) is thought to play a key role in the process of photosynthesis that is used by plants to capture energy from the Sun and good experimental evidence for this behaviour is available based on ultra-fast spectroscopy with pulsed lasers. Photosynthesis is a crucial process in renewable energy technologies, in addition to its importance in global carbon dioxide sequestration and food production. The way that biology manages to manipulate quantum coherence phenomena at room temperature may provide researchers with some lessons into fundamental physics. To date quantum coherence effects with electrons only seem to be measurable in fairly esoteric systems that weakly interact at very low temperature, e.g. in single atom traps or with superconductors. These extreme conditions, although of fundamental interest, mean that useful applications are much more challenging. Specifically, it is the control of sample/environment interactions that biology seems to do well, which would otherwise scramble the coherence over a negligibly small time scale, i.e. at high temperatures, high concentrations and with strong interactions. These coherence effects are why Schrödinger's famous thought experiment on a cat that is simultaneously neither alive nor dead cannot be yet performed in reality.

Additional quantum optics effects occur in human vision, where the detection limit for photons can approach that of the single photon level. The dopsin molecules that detect light in human eyes have demonstrated single photon sensitivity *in vitro* experiments, but this level of sensitivity can be degraded by an order of magnitude by noise in the biochemical amplification circuits in living people. Nevertheless, the quantum nature (chunkiness) of light can still directly relate to people's experience of low light imaging with their own eyes.

Currently a hotly contested field is that of the quantum mechanics of magnetism used for navigation in birds. The entanglement of electrons in molecules in the eyes of some birds is thought to provide them with an inbuilt compass to guide their migrations. Experiments are tricky in this case, because subtle quantum mechanical phenomena need to be interpreted in the presence of the intricate behavioural responses of birds. The field thus requires the development of quantum mechanical aviaries to explore a real *Schrödinger's duck*, rather than a hypothetical *Schrödinger's cat* that forms a classic thought experiment to understand entanglement in the field of quantum mechanics.

In the general case, the many-body problem of electrons in molecules cannot be solved exactly when described with Schrödinger's equation (chapter 17) and this is the main theoretical tool that is used to describe the quantum mechanics of molecules. This lack of a solution is not specifically a problem with the Schrödinger equation and it appears to be intrinsic to all realistic descriptions of the many-body problem with a long-range interaction. Pragmatically the method of phenomenological bootstrapping is needed as described in chapter 17, since the many-body problem is in general intractable.

There were some slightly overambitious comments made by Paul Dirac (one of the fathers of quantum theory and a Nobel laureate) after the development of an early version of quantum mechanics in the late 1920s, that chemistry was now entirely understood. The semantics of quantum field theory for elementary particle phenomena is now relatively well developed (the extension of quantum mechanics to describe all matter), but many tools to make the theory appropriate for calculations with all molecules do not exist and predictions are of a low quality in most practically relevant systems.

There is clearly lots of important work still left to be done to provide biology with some solid quantum mechanical foundations. However, such quantum mechanical behaviour ultimately needs to be captured to provide high resolution descriptions of all biological physics problems.

14.2 Suggested reading

Introductory

Al-Khali J and McFadden J 2015 *Life on the Edge: The Coming of Age of Quantum Biology* (London: Black Swan). A good popular account of modern ideas for the application of quantum mechanics in biology, although the later sections of the book are a little speculative.

Ball P 2000 *H2O: A Biography of Water* (London: W&N). Water is given the spotlight in this popular account that discusses many of its anomalous properties.

Advanced

Feynman R 2011 *Feynman Lectures on Physics, Volume 3* (New York: Basic Books). Entanglement and tunnelling are much better described using the correct mathematical tools. This book is a good place to start to learn the fundamental concepts that are necessary for a detailed understanding of quantum mechanics, and is provided by an innovator in the area of quantum field theory.

Bialek W 2012 *Biophysics: Searching for Principles* (Princeton, NJ: Princeton University Press). A detailed discussion of the quantum mechanics of vision is presented in this advanced textbook from a leader in the field.

Comez L, Paolantoni M, Sassi P, Morresi A and Fioretto D 2016 Molecular properties of aqueous solutions: a focus on the collective dynamics of hydration water *Soft Matter* **12** 5501–14. A recent review of experiments and theory that relate to the collective properties of water molecules measured using ultra-fast spectroscopy techniques.

Lambert N, Chen Y N, Cheng Y C, Li C M, Chen G Y and Nori F 2013 Quantum biology *Nature physics* **9** 10–8. Reviews some phenomena in biology in which quantum phenomena are thought to play a significant role.

Levine I N 2013 *Quantum Chemistry* (Upper Saddle River, NJ: Prentice Hall). A classic undergraduate text on quantum chemistry.

Piela L 2014 *Ideas of Quantum Chemistry* (Oxford: Elsevier). An introduction to quantum chemistry for undergraduates.

14.3 Technical details

14.3.1 Anharmonic potentials

To describe a quantised interaction between particles it is useful to first define a potential that describes their interaction. A simple harmonic potential ($V(x)$) in one spatial dimension is described by the equation

$$V(x) = \frac{1}{2}k(x - x_0)^2 \tag{14.1}$$

where x is the position in one-dimension, k is the elastic constant and x_0 is the position of the minimum of the well. The equation describes the energy stored in a Hookean spring. Mathematically equation (14.1) is a good approximation for any potential well (figure 14.1), if the displacements from the equilibrium position are very small. However, in a wide range of applications this requirement is not obeyed and more sophisticated potentials are needed. In the molecular physics community the Morse potential is often used as an alternative, as it is an anharmonic potential and makes the resultant calculations slightly more realistic (Banwell and McCash 1994). The equation describing the Morse potential is

$$V(x) = D_e(1 - e^{-a(x-x_0)^2}) \tag{14.2}$$

where D_e is the depth of the potential, a is a characteristic constant and x_0 is the position of the minimum. The advantages of the Morse potential are that it leads to an analytic solution of the Schrödinger equation, predicts the convergence of energy levels at large inter-nuclear distances and describes the phenomenon of molecular dissociation (the last two attributes are not found with harmonic potentials). Equation (14.2) is a convenient empirical expression, but it lacks a fundamental physical motivation. Ideally a derivation based on the underlying quantum electrodynamics is required.

Rather than simple harmonic motion in a harmonic potential ($m\ddot{x} + kx = 0$, with a potential given by equation (14.1)), it is also possible to consider a fractional harmonic oscillator

$$\left(m\frac{\mathrm{d}^{2\alpha}}{\mathrm{d}t^{2\alpha}} + k\right)x = 0 \tag{14.3}$$

where α is a positive exponent. This provides a new, relatively unexplored method to understand oscillatory systems. It is also possible to write down a fractional Schrödinger equation to describe quantum mechanical systems (Hernmann 2011).

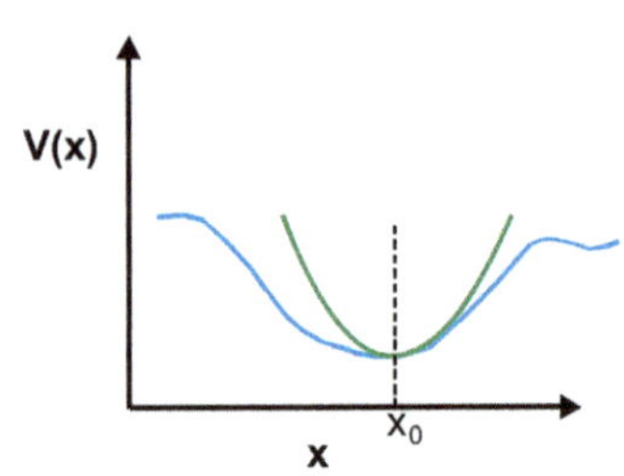

Figure 14.1. The potential ($V(x)$) as a function of displacement in one dimension (x) inside a molecule. x_0 is the equilibrium separation. A general anharmonic potential (blue) and a harmonic approximation (green) used to describe the interactions in a molecule.

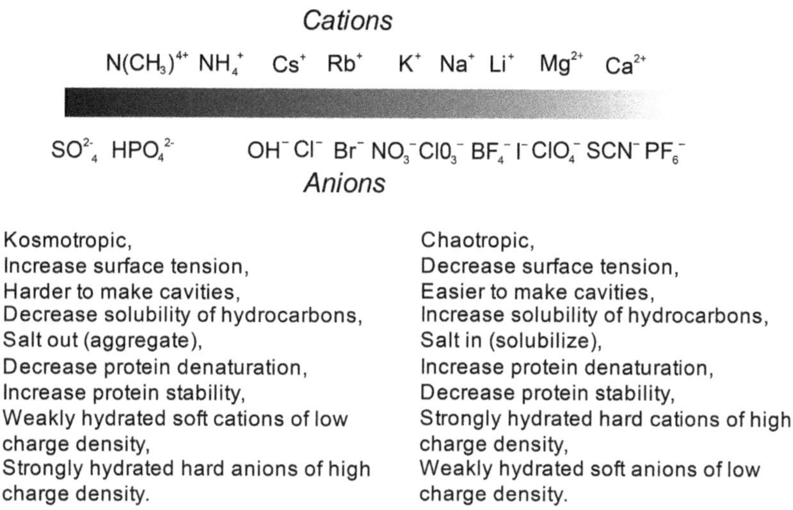

Figure 14.2. Typical ordering of cations and anions in a Hofmeister series (Kunz 2010).

14.3.2 Hoffmeister series

The Hofmeister series has lacked a convincing quantitative explanation, since its discovery over 120 years ago. The series can play an important role in a vast range of biological systems, since they all involve salty solutions. Some recent progress towards a quantitative description has been made in terms of ion affinities (figure 14.2) (Kunz 2010).

14.3.3 Hydrogen bond potential

The simplest mathematical approximation for the potential due to hydrogen bonds (involving polarised molecular bonds) is to consider them as point charge/dipolar interactions (Israelachvili 2011). The classical prediction for the potential ($w(r)$) as a function of distance (r) is given by

$$w(r) = \frac{-Q_H u \cos\theta}{4\pi\varepsilon_0 \varepsilon r^2} \tag{14.4}$$

where Q_H is the magnitude of the point charge, u is magnitude of the dipole, θ is the angle between the dipole and the line to the point charge, ε is the relative permittivity and ε_0 is the permittivity of free space. This is just a first approximation and does not properly capture either the magnitude or the directional effects of the hydrogen bonds.

Furthermore, entropic effects also need to be considered in the calculation of potentials due to hydrogen bonds that give rise to the *hydrophobic effect*, since the motion of water molecules becomes constrained around molecules that they are not able to interact strongly with, i.e. water becomes structured around hydrophobic molecules.

To introduce detailed quantum mechanical phenomena into this framework is a tricky area, but some useful progress has been made. Up-to-date coverage can be found in *Quantum Chemistry* (Levine 2013).

References

Banwell C and McCash E 1994 *Fundamentals for Molecular Spectroscopy* (New York: McGraw-Hill)

Hernmann R 2011 *Fractional Calculus: An Introduction for Physicists* (Singapore: World Scientific)

Israelachvili J 2011 *Intermolecular and Surface Forces* (London: Academic)

Kunz W 2010 Specific ion effects in colloidal and biological systems *Current Opinion in Colloid and Interface Science* **15** 34–9

Levine I N 2013 *Quantum Chemistry* (Upper Saddle River, NJ: Prentice Hall)

Chapter 15

The structure of membranes—uncharted factories

15.1 Introduction

The field of structural biology is far from complete, although it sometimes can seem a little old fashioned and stodgy to many of its practitioners. There are a number of factors that make structural biology appear slightly drab and unappealing to modern biophysicists. The major one is that *molecular biology* is now around 50 years old, whereas *structural biology* is even older at a venerable 95 years of age. The terminology, *molecular biology*, was introduced by W Astbury in a *Nature* article written in 1961. X-ray experiments to understand the structures of fatty acids found in biological membranes were first performed in the early 1920s. The fields of molecular and structural biology have thus become so well established that it is easy to lose sight of the many important outstanding questions that can attract the attention of new generations of researchers hoping to make a name for themselves.

In principle, the rules of the game in structural biology are relatively straightforward and predictable. Take a molecule, or an aggregate that contains a well-defined number of molecular components, and try to fully define its architecture, preferably with an atomic level of resolution. People often think about frozen structures, where all the atoms are held in fixed relative positions. Nowadays, three-dimensional printers can be used to create solid macroscopic versions of the molecular structures from plastic, to use them as exotic paper weights or, more usefully, to see how the molecular structures complement one another. However, these frozen pictures are always an extreme approximation, since thermal fluctuations constantly move all molecules around in space at small length scales. It is as if the molecules are caught in a hurricane and they continually jiggle about due to the intermittent gusts of thermal energy in their environment. The frozen molecular images that appear in textbooks tend to have been historically led, since crystallography techniques only highlight the averaged positions of the static components of crystalline samples, and

doi:10.1088/978-0-7503-1374-2ch15

© IOP Publishing Ltd 2017

highly mobile atomic motions are relatively invisible in the standard x-ray techniques used. Furthermore, thermal motions of atoms are much smaller when they are constrained inside crystals, and the crystalline structures are thus not perfectly indicative of the molecular structure in living organisms, which tend to be much less tightly constrained. The majority of molecules form fluid-like phases in cells. Very few truly crystalline molecules actually occur in living organisms. A few exceptions are gout in human joints (uric acid crystals) and some lipids held in the seeds of plants for long term storage. Life as a rule prefers to use dynamic flexible states of molecules.

The vast majority of biological molecules thus contain fairly flexible arrangements of atoms. Nevertheless, the rigid structures calculated from crystallography experiments do present useful reference architectures that provide analogues to help to begin more extensive research on their flexible relatives. It is therefore extremely inconvenient that biological membranes experience large undulations at room temperature and predominantly do not crystallise under any conditions. This makes x-ray crystallography, and thus the field of structural biology of membranes, much more challenging. Therefore, even rigid reference models can often not be created. Furthermore, membranes are fairly disordered and fluid-like (historically this was called the *fluid mosaic model*), which means enumeration of all the different atomic positions within the membrane is an impossible task. In general, there are no simple crystalline lattices (wall papering patterns at the atomic scale) that can be used to describe the structure of a membrane. The undulations of the membrane mean unusual interactions occur when membranes collide, which can lead to a repulsive force. Hydrodynamics also play a role since the surrounding fluid must move around the membranes, modifying the time scales at which the undulation forces are experienced. Good quantitative models exist for undulation forces in synthetic membranes that have been verified by direct measurement (e.g. using surface force apparatus), but measuring these forces in living biological cells is much more tricky.

Research groups have solved some of the problems with respect to the nanoscale organisation of membranes in the cell with new microscopy methods and theoretical tools, but there are still numerous large gaps in our understanding. For example, the structures adopted by membranes are no longer thought to be a uniform blend of their components. Small micro-islands float around on membranes that are enriched in specific molecular components called *lipid rafts*. For example, super-resolution fluorescence microscopy experiments from our own group indicate that the Rab5 proteins discussed in chapter 1 used in signalling with endosomes appear to exist in lipid rafts. The exact function of these rafts is still not well understood (physically they are created by a process of liquid–liquid phase separation in two dimensions, similar to the processes involved in vinaigrette recipes that mix oil and water in three dimensions) and a quantitative understanding is still absent.

Many of the major factories inside the cell that process biological molecules are built with membranes and the majority of molecules synthesised inside cells travel through two highly membranous regions when they are processed. These molecular factories are called the endoplasmic reticulum (the ER, step one) and the Golgi apparatus (step two). The two-dimensional membranes in these two regions are thought to help them manufacture biochemicals (proteins, lipids, carbohydrates etc) at a faster rate, since reactions often go faster when constrained to two dimensions rather than three (it is a statistical effect due to the improved probability that reactants meet their substrates in two dimensions). Many question marks still hang over the structure and the function of both the membranes in the ER and the Golgi apparatus.

L. *Question:* Determine accurate functional maps for the architectures of the endoplasmic reticulum and the Golgi apparatus, and demonstrate whether they are arranged in a random higgledy-piggledy fashion, i.e. discover the functionality of their designs.

It seems hard to believe that such important assemblies as the ER and the Golgi apparatus are not carefully constructed and maintained inside cells. Yet, when the research literature is examined for an accurate functional picture of these vital organelles, only higgledy-piggledy messes are seen. Lots more detail is known about the structure of the ER, such as the sub-classifications of *rough* (studded with ribosomes, nanomachines that make proteins) and *smooth* (no ribosomes, but lots of fat molecules) regions. Nevertheless, the underlying organisational principles are still not clear from a physical perspective and the current model is messy. In part this is due to the extreme heterogeneity of the membranes' chemistry. Specific proteins in the membranes can be labelled at the level of single molecules using fluorescence microscopy, but images quickly start to look like kaleidoscopic patchwork quilts. Thousands of different labels would be needed to simultaneously locate all the protein interactions in the ER and Golgi apparatus, let alone all of the different lipids and carbohydrates they manufacture. The implicit assumption for such fluorescence microscopy experiments is that the inclusion of all the (often bulky) fluorescent labels does not perturb the structures that are probed and this is not necessarily the case. Furthermore, researchers are lucky if they can label a few different components (amounting to single digits) using current fluorophore technologies.

The analogy can be made of the ER to a busy city at rush hour. Scaling up the globular proteins (the main traffic in the ER) to the size of a car, a map of the city on the size of around 25 square kilometres is needed for the ER. Moreover, it is a labyrinthine city arranged in three dimensions, where transport occurs on a variety of height levels. Ideally all the transport processes should be probed to provide a zoom facility into different regions with improved resolution, similar to traffic reports made with helicopters for daytime television that can zoom into a particular road that contains a traffic jam. Transport at the street level can then be rigorously connected with the global dynamics of the city using quantitative theoretical tools.

A comparison of the ER's architecture can also be made with the microstructure of chromosomes in the cell's nucleus. Nuclear chromosomes are not fixed in space, so they provide flexible mobility for their components, but their architecture is now known to be carefully positioned. Each chromosome has its own well-defined neighbourhood and access to the different sections of the DNA chains is precisely regulated. Without a well-defined chromosomal ultrastructure, the wrong genes would be switched on and off, and hence the wrong gene circuits activated, with catastrophic consequences for the organism. Catastrophic problems would also be expected for the molecular assembly lines in the ER and the Golgi apparatus, if they were not carefully controlled, although the mechanism through which control is achieved may be completely different to that of chromosomes.

The extreme chemical heterogeneity of the ER and the Golgi apparatus is mirrored in the diverse range of roles that the structures can perform, e.g. calcium is stored in the ER and a wide range of proteins (the ER), steroids (the ER), lipids (the ER and the Golgi) and carbohydrates (the Golgi) are synthesised and sorted there.

Recent breakthroughs in the structure of the ER have come from super-resolution fluorescence microscopy and tomographic electron microscopy experiments. The old-fashioned picture of stacks of membranes solely arranged in lamellae (reproduced in a huge number of textbooks) is wrong and it is an artefact of the low time and the low spatial resolution of the microscopy images used to create the picture. Many of the membrane stacks are actually arranged in tubules, whose diameter (on the order of a hundred nanometres) is defined by membrane proteins that adopt wedge shapes that are embedded along the length of the tubes. However, the exact significance of the tubules is not well determined, and therefore neither is their functional role. An additional complication is that the tubules are extremely soft and exhibit large amplitude vibrations, blurring attempts to image them. A large challenge in the field is therefore to:

L. *Question:* Determine the physical mechanisms through which tubules in the endoplasmic reticulum function.

Experimentally one of the foremost techniques to follow membrane structures in cells is cryotransmission electron microscopy. The membranes are frozen very quickly when they are plunged into liquid ethane (−188 °C) and it is hoped that the freezing process occurs faster than that of any structural modifications, which will disrupt the native architecture. The samples are then cut into thin slices, placed into an ultra-high vacuum and bombarded with electrons in the microscope. The resultant images of membranes in the ER have been made with nanometre resolution. Researchers have made rapid process in this field and a current emphasis is to use image analysis in three dimensions to interpret the complex topologies created by membranes. Ideas from differential geometry are required to interpret the data, such as the Gaussian curvature of the membranes. *Differential geometry* is a key mathematical discipline used to describe the curvature of space–time in general relativity (Riemann curvature). The curvature of membranes in biology can be described using similar mathematical tools (Gaussian curvature) to understand the structures they adopt, e.g. the morphology of endosomes as they interact with the endoplasmic reticulum. Frozen structures do, however, only tell half the story. Fluorescence microscopy images from a colleague's laboratory (Allan) show the tubules in the ER have large amplitude fluctuations driven by thermal energy (they are soft structures) and the action of motor proteins may also contribute to their motion (facilitating the mixing of reactants).

In addition to a functional map for the endoplasmic reticulum, it is important to understand the dynamics of the assembly lines that are in action in these molecular factories.

L. *Question:* Create a general map for the assembly lines and sorting processes that exist in the endoplasmic reticulum.

Such a map of the ER assembly lines would allow the creation of a bird's eye view of the processes involved in this crucial organelle, analogous to that of pictures taken by a helicopter of traffic in a district of a city during the rush hour. A major challenge in such a study is to integrate the interactions of all the different components into a single coherent picture, i.e. create an integrative model of the ER (and then the Golgi apparatus). Eventually this could provide unparalleled insights into how biological molecules are created inside cells and many of the diseases associated with their malfunction, such as diabetes, haemophilia, motor neuron disease etc.

15.2 Suggested reading

Introductory

Buehler L K 2016 *Cell Membranes* (New York: Garland). An up-to-date introduction to the biochemistry of membranes.

Moritsen O G 2004 *Life - as a Matter of Fat: The Emerging Science of Lipidomics* (Berlin: Springer). A popular account of membrane research.

Advanced

Philips R 2013 *Physical Biology of the Cell* (New York: Garland). An instructive introduction to the physics of membranes.

Boal D 2012 *Mechanics of the Cell* (Cambridge: Cambridge University Press). Another good introduction to the physics of membranes.

Hu J, Shibata Y, Shemesh T, Li Z, Coughlin M, Kozlov M M, Rapoport T A and Prinz W A 2008 Membrane proteins of the endoplasmic reticulum induce high-curvature tubules *Science* **319** 1247–50. Primary research literature that describes some of the membrane proteins that determine the structure of tubules in the endoplasmic reticulum.

Nixon-Abell J *et al* 2016 Increased spatiotemporal resolution reveals highly dynamic dense tubular matrices in the peripheral ER *Science* **354 aaf3928.** During the preparation of this manuscript a fundamental contribution was made to this issue of tubules in the endoplasmic reticulum by Eric Betzig (Nobel Prize winner for chemistry in 2014) and collaborators. A current challenge is to build on this study.

15.3 Technical details

The old-fashioned picture of the endoplasmic reticulum seen in the schematic diagram of a cell shown in figure 15.1 is very vague and it is the type of diagram students are required to memorize for secondary school exams. The morphology of the endoplasmic reticulum (shown as a compact grey object near to the nucleus and it was thought to be full of layers of membranes) is now known to be completely wrong. Many of the continuous membranes shown in such textbook schematics are an artefact of the low time and space resolution of the previous generation of fluorescence microscopes used to image them. Furthermore, the ER can stretch all the way across cells (figure 15.2).

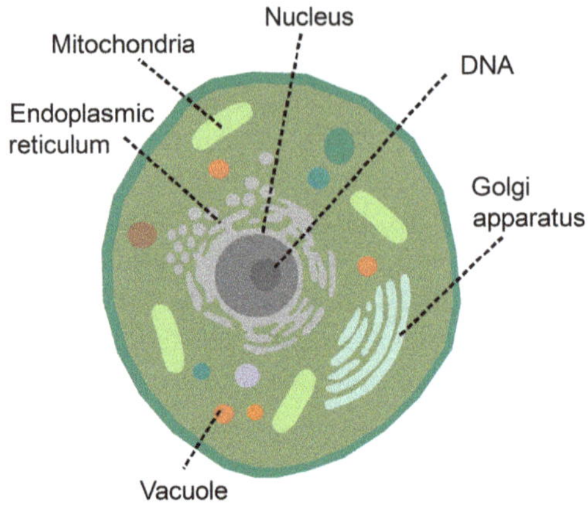

Figure 15.1. Old-fashioned schematic picture of a eukaryotic cell. The endoplasmic reticulum (ER, light grey) has a layered structure close to the nucleus. In reality the ER is known to have a much more extensive structure (figure 15.2) and it is primarily formed from tubules.

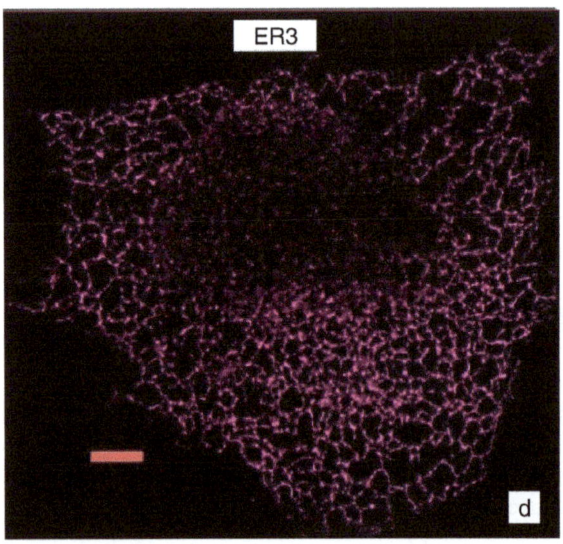

Figure 15.2. Super-resolution fluorescence microscopy image of the endoplasmic reticulum with a resolution of around 30 nm in a human lung cell. A network of membrane tubules is visible that are ~90 nm in diameter. Reproduced with permission from Georgiadis *et al* 2017.

15.2.1 Fluctuations of vibrating tubules

The vibration of strings is a standard topic in undergraduate physics courses and the standard models that are taught can be used to describe the motion of violin strings in air. The vibrations of fibres in water are slightly more complicated due to their strong hydrodynamic interactions, but some progress has been made to find analytic expressions to describe the motions of single fibres, e.g.

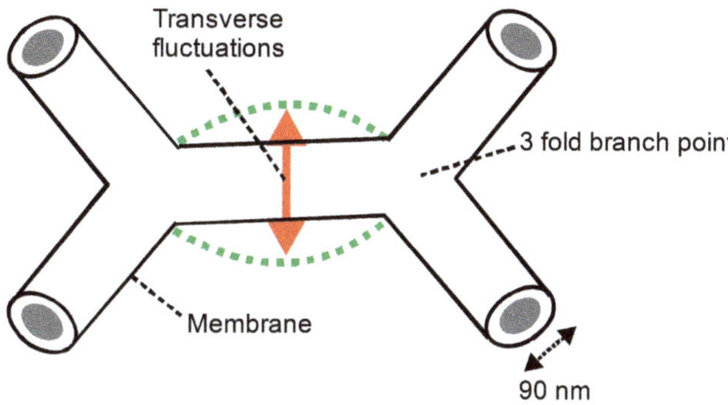

Figure 15.3. The transverse fluctuations of a segment of a tubule in the endoplasmic reticulum. The branches of the network all tend to have a three-fold connectivity.

$$\langle \Delta r_{\perp}^{2} \rangle = 0.47 \left(\frac{k_{B}T}{\eta} \right)^{3/4} \frac{1}{L_{P}^{1/4}} t^{3/4}$$

where $\langle \Delta r_{\perp}^{2} \rangle$ is the mean square displacement of the transverse fluctuations of the fibres, kT is the thermal energy, η is the viscosity of water, L_p is the persistence length and t is the time (Granek 1997). Calculations become trickier when the collective motion of semi-flexible fibres (such as the ER tubules) is considered, but again good progress has been made, e.g. reptation of fibres due to their snake-like motion in tubes (Broedersz and MacKintosh 2014). However, the fibres in the endoplasmic reticulum have a hexagonal network morphology, so additional calculations are needed to understand the impact of the geometry on the elastic response of the full network. The ER tubules are also very soft and the nanomotors in their environment may be driving some of the fluctuations (figure 15.3). How they do this in detail is not well understood, but some calculations are now available for actively driven semi-flexible fibres (Ghosh and Gov 2014).

References

Broedersz C P and MacKintosh F C 2014 Modelling semiflexible polymer networks *Rev. Mod. Phys.* **86** 995–1033

Georgiadis P, Allan V, Wright G D, Woodman P, Udomai P, Chung A and Waigh T A 2017 How soft are the tubules in the endoplasmic reticulum? Submitted

Ghosh A and Gov N S 2014 Dynamics of active semi-flexible polymers *Biophys. J.* **107** 1065–73

Granek R 1997 From semi-flexible polymers to membranes: anomalous diffusion and reptation *J. Phys. II France* **7** 1761

IOP Publishing

Chapter 16

Drug delivery—gene therapy and other stories

16.1 Introduction

Gene therapy, similar to stem cell treatments, is heralded as one of modern medicine's most powerful strategies to combat disease and it can in principle be used to cure a wide variety of different medical conditions. Many of these conditions were until recently thought to be completely incurable. The principle is simple, just find the location of a faulty gene (diseases such as cystic fibrosis and haemophilia are thus prime targets, since they are often associated with point mutations that involve a small number of faulty genes) and replace it with a functional equivalent. Mechanistically one of the main challenges to this process of gene therapy is thus to transfer the replacement DNA into the nucleus of the target cell. This replacement DNA needs to travel through the cell membrane, avoid degradation in the cytoplasm, travel to the nucleus, pass through the nuclear membrane and replace the faulty gene(s) in exactly the right place in exactly the right chromosome (ideally the faulty sections should also be excised from the genome). Sometimes the process of gene replacement needs to be repeated multiple times if the genome contains multiple copies of a malfunctioning gene.

Until very recently man's attempts to modify specific sections of DNA in humans has been fairly haphazard. Genes could be replaced, but it was tricky to do so in exact regions of a chromosome in a human cell. Expression of a foreign gene anywhere in a chromosome was counted as a big success. However, genetic modification has been recently revolutionized by the introduction of the CRISPR technology, which is a molecular technology originally found in the immune system of bacteria. CRISPR provides a molecular machine (Cas 9) that can cut a precise region of DNA that corresponds to a matching guide strand of RNA. New sections of DNA can then be recombined with the cut strands to create a final double helix where, for example, the faults that cause a disease have been removed. The CRISPR mechanism works fairly well in a test tube, but in common with previous mechanisms for gene treatments, its delivery into patients has been problematic. Currently the most realistic strategy for

doi:10.1088/978-0-7503-1374-2ch16 © IOP Publishing Ltd 2017

gene therapy with CRISPR is to modify the cells outside the body and then to deliver the cells themselves to the damaged regions.

In contrast with the large improvements achieved with *in vitro* gene editing techniques, progress in the creation of vectors to transmit genetic tools to the nucleus of a cell in a living organism have been relatively slow. In general, huge challenges are presented to a successful drug molecule, such as a gene vector, for it to be effectively used in a living organism. Typically, the drug needs to pass through a mucus barrier (that exists across the lungs, the mouth, the eyes, the intestines, the stomach, etc), be absorbed through a lining of skin cells, travel to the blood stream and finally release its cargo in the correct position in the organism, so that it is received by the correct variety of cells. Furthermore, the gene vector needs to arrive without triggering the organism's immune system or disturbing any other of the vital metabolic processes that are in action. Intravenous injection is of course possible and this circumvents two of these initial barriers to drug delivery, but it is slightly more invasive and correct loading into the blood stream still has some challenges. In general, the processes are so complicated during drug delivery and the range of possible side effects so vast, that once a lead compound has been found by chemists, a wide range of animal and clinical testing is needed before the drug can be defined as relatively safe and be thought of as an effective treatment. It is expected that this will continue to be true in the 21st century for the foreseeable future.

Many qualitative physical phenomena have been identified that describe crucial steps in drug delivery. These include a wide variety of electrostatic interactions in gene therapy, since DNA molecules are conveniently transferred when they are complexed with oppositely charged molecular vehicles, e.g. the negatively charged DNA chains can stick to positively charged polyethylene imine chains. The exact forces between the charged molecules can cause dramatic variations in the structure of the complexes that are formed and exact quantitative expressions that describe the forces are still open for debate (chapter 17). Complexation of DNA can be considered as a phase transition.

Self-assembly describes the mechanism with which molecules can spontaneously organise themselves into well-defined structures in solution, e.g. the lipids that aggregate to form the membranes of cells or the proteins that tessellate the coats of viruses both self-assemble. This apparently magical process is now known to be due to simple thermodynamic effects (a balance between the entropy and the internal energy) and good quantitative models exist, originally developed for the detergent (soap) industry. Soap molecules self-assemble around dirt molecules to stabilise them during cleaning processes and the soap can also lever off dirt molecules from surfaces using a depletion interaction. Self-assembly for gene vehicles, such as liposomes and viral gene carriers, are important mechanisms in gene therapy, where the process can create well-defined structures for nanovehicles that can then be used to transfer DNA. Once the processes of electrostatic complexation and self-assembly are better understood the suggestion is:

L. *Question:* Create a new more effective vector for gene therapy based on the quantitative physical principles involved.

As soon as accurate physical models are available for gene vectors the process of their design and synthesis can be iterated and thus optimised, e.g. the size, the shape and the charge adopted by a gene vector can be accurately engineered.

Surprisingly, simple molecular systems have been effective to transfer genes into the nucleus of cultured human cells, such as polyethylene imine (a simple synthetic polymer, similar to polyethylene that is positively charged), liposomes (positively charged lipids) and polysomes (positively charged synthetic polymers that are surface active and self-assemble in a similar manner to lipids). These systems can successfully transfer DNA fragments into a cell and they are subsequently expressed by the genetic machinery inside the cell, e.g. a fluorescent protein can be used as an indicator when the transferred DNA is genetically expressed in the target nucleus, which is easily seen under an optical microscope. However, the creation of completely new chemistries for gene therapy is probably more of a job for a medicinal chemist than a biophysicist. A more relevant biophysical question is:

L. *Question:* Create an accurate quantitative model for the rate of transcription and expression of DNA combined with a gene vector.

This is a challenging problem, since gene control is still not sufficiently well understood (chapter 11), with multiple complementary mechanisms that all play key roles and are as yet not well defined, such as the histone code, regulatory RNA chains and the looping of the DNA chains. Nevertheless, simple kinetic models should help to improve experimental parameters that determine the rate at which DNA is delivered in gene therapy and then transcribed. Fast transcription often implies the process of gene therapy will be more effective. A crucial quantitative statistical model describes the rate of transcription of a DNA chain by the standard proteins that is required to express it (RNAP). It is much faster when the RNAP associates with the DNA chain and moves along it in one dimension to search for a sequence to transcribe, rather than searching in three dimensions. The model is in reasonable agreement with experiment, but many important biological details need to be properly included (e.g. gene regulatory proteins such as promoters or inhibitors) and sub-diffusive motions due to the congested environment may also play a role (chapter 12).

Currently the best vehicles (vectors) for gene delivery are provided by viruses. All viruses are naturally evolved vehicles for gene delivery that allow these parasites to proliferate. They infect cells and then deliver their payloads of genetic material into the cells' nuclei; these in turn hijack the cells' own molecular machinery to make more copies of themselves. This process of viral infection often leads to the deaths of cells and the dead cells burst open to release more copies of the virus that can then go on to infect neighbouring cells. The principal medically relevant viruses that infect human cells, such as Rabies, are exquisitely well designed. The viruses can even hijack motor proteins inside the cells to provide faster delivery to the nucleus, to increase their rate of gene delivery and thus their degree of virulence. Rabies does a much better job of gene transfer than is possible with any of the synthetic gene vectors that are currently available. Clearly viruses need to be better investigated to understand how they achieve this feat. Millions of years of evolution have made viruses the most effective vehicles to deliver genes that currently exist. Fundamental research into viruses also has a clear driver to provide an understanding of their role in disease and their important roles in determining the populations of micro-organisms in all ecosystems on planet Earth (viruses also infect microbes). Viruses that specifically target bacteria (bacteriophages) are also good candidates for antibacterial treatments, since they can be accurately targeted at specific bugs (chapter 10).

A few viruses have an even more effective strategy for gene transfer in that they insert their DNA into a cell's genome to make a permanent genetic modification. The cells infected by these so called *retroviruses* will create more viruses in perpetuity and, if the cells divide, the daughter cells will also be infected. The human body has no effective mechanism to remove this type of infection. Examples of such retroviruses include HIV, SIV and leukemia virus. Carefully attenuated versions of such viruses are now standard molecular tools in laboratories to make stable genetic constructs inside cells, but effective treatment of the diseases associated with them is still challenging.

Thus, given the choice to synthesise a gene vector from scratch or hijack a naturally occurring vector (a virus), it seems much simpler to borrow the design of a virus. However, immune responses to viral carriers are a major issue, which have produced fatalities in initial trial experiments of gene therapy with viruses. The immune system in humans is there to combat infections and it is often not able to differentiate between a harmless virus used as a gene vector and a more malignant variety that will cause a disease. DNA has been successfully transfected into patients using viral gene therapy, but some people have died from unrelated side effects (specifically acute immune responses). It is hoped that this extremely effective mechanism of viral gene therapy can be made more successful in humans in the near future, and its prospects mainly depend on whether a better quantitative understanding of how the human immune system functions can be created.

Quantitative research into gene therapy has also been partially blocked by an incomplete understanding of the processes through which nanoparticles enter cells (*endocytosis*). Endocytosis is an extremely important process in determining how cells exchange molecules with their environment and it is a crucial mechanism involved in molecular communication and thus information transfer inside organisms, and as such it was briefly described in chapter 1.

L. *Question:* Create an accurate quantitative model for the endocytosis of gene vectors into human cells.

Endocytosis is biologically a rich and varied process, and developments in clathrin coated endocytosis were the subject of a Nobel Prize in Medicine in 2014. However, there are still many gaps in the picture of endocytosis from even a qualitative biochemical perspective, e.g. how do non-clathrin coated endosomes function? Physically, membranes can be considered as elastic sheets and energy is stored in them when they are deformed into spherical bags, i.e. endosomes. A combination of motor proteins is needed (to pinch off the spherical bags) with membrane proteins (often pie shaped membrane proteins do the job) to encourage the curvature of the endosomes. A complex dynamic range of different morphologies is possible with endosomes driven by force laws at the nanoscale, e.g. electrostatics, van der Waals, steric forces and hydrogen bonds can play a role.

Another challenge in the biophysics of drug delivery is presented by mechanisms of drug formulation. The active component of a drug is often a minority species in the mixtures that are bought from a pharmacist (called the *formulation* of the drug), whether it is an antibody for the treatment of a cancer or a viral inoculation for gene therapy. Often additional molecular components are needed to improve the active component's solubility, viscoelasticity, taste, passage through the stomach, etc. A number of headaches are therefore presented to researchers on issues of phase stability (e.g. to ensure that all the molecules do not stick together to form a random inactive mess), so that the active component remains effective, but also that the syringability of the molecules is maintained, i.e. the viscoelasticity of the mixture is not so high that injections can no longer be performed. Painful needles that have large bores are required for formulations with low syringability and they can seriously impair the practicality of a particular drug.

Issues with high viscosity formulations are not always relevant for drug delivery. For example, only ultra-low concentrations of Botox are needed in this extreme case due to the highly poisonous nature of the drug. Botulism is a disease caused by the same molecule that is in Botox and it can quickly lead to death, since Botox is a potent neurotoxin. Thankfully, in the modern world botulism poisoning is fairly rare and it is most often caused by bacterial infections due to improperly performed canning procedures in DIY attempts to preserve home grown vegetables. In cosmetic surgery Botox is used in procedures to remove wrinkles through the partial paralysis of muscles in the skin. Nevertheless, this is an exception to the rule and many modern drugs require high concentrations to improve their efficacy.

One of the modern paradigms for anti-cancer treatment is based on antibodies (*immunotherapy*). These are extremely selective molecular glues and they can be specifically targeted at cancerous cells. Thus, one challenge is to provide optimal formulations of such molecules to form effective treatments.

L. *Question:* Modify the formulation of globular proteins in concentrated solutions (e.g. antibodies) to reduce their viscoelasticity, whilst maintaining the activity of the active components in these solutions.

Blockages can form in a similar way to traffic jams on a busy road. Large micron sized particles can spontaneously form load bearing arches in the flows that cause the jams. Whether this happens for blockage with flowing nanoparticles (such as globular proteins) on length scales a hundred times smaller is open for debate. Syringes are often coated with a thin layer of oil to reduce the frictional effects on the boundaries, but how this affects the formation of arches is not well understood. The oil can also detrimentally affect the proteins by unfolding them into inactive conformations.

Similar questions on the compositions of the formulations of drug molecules arise for the viscoelasticity of suspensions of nanoparticles used for gene therapy, such as viruses and cationic polymers, e.g. gene vectors made with complexes of polyethylene imine and DNA. Often it is advantageous to work with high concentrations of the active component to maximise the dosage, but they become highly viscoelastic and thus hard to deliver.

Historically, drug delivery has been considered by the disciplines of pharmaceutical science and medicinal chemistry. It is hoped that biological physics can provide the field with a novel quantitative perspective to help such revolutionary techniques in modern medicinal research as gene therapy and immunotherapy.

16.2 Suggested reading

Introductory

Calladine C R and Drew H 2004 *Understanding DNA: The Molecule and How it Works* (Cambridge: Academic Press) Classic introduction to the properties of DNA from an engineering perspective.

Florence A T and Attwood D 2015 *Physicochemical Principles of Pharmacy: In Manufacture, Formulation and Clinical Use* (London: Pharmaceutical Press) A classic text that describes current methods to understand the physical principles of drug delivery.

Patrick G L 2013 *Introduction to Medicinal Chemistry* (Oxford: Oxford University Press) A wide-ranging tour of drug molecules is provided, combined with some practical issues needed to take the molecules to the market place.

Rang H P 2015 *Rang and Dale's Pharmacology* (London: Churchill Livingstone) Considers the impact of drug molecules on the physiology of living organisms.

Advanced

Waigh T A 2014 *The Physics of Living processes: A Mesoscopic Approach* (Chichester: Wiley) Chapter 18 'Phase behaviour of DNA'. Considers some of the physics involved in gene therapy and gene expression, such as the questions of polyelectrolyte complexation and self-assembly.

Parham P 2014 *The Immune System* (New York: Garland) The extremely complex activity of the human immune system needs to be understood to increase the efficiency of viral gene transfer. This book provides a simple introduction to qualitative biochemical aspects of the field.

Chaudhuri A, Battaglia G and Golestanian R 2011 The effect of interactions on the cellular uptake of nanoparticles *Phys. Biol.* **8** 046002. A quantitative understanding of endocytosis is still at an early stage of development, but this research provides a useful reference point.

16.3 Technical details

16.3.1 Gene therapy

A challenge in gene therapy is to understand the kinetics of the molecular processes involved (figure 10.1). Once naked DNA chains are transmitted into a human cell they can be very quickly broken down, so the challenge is to rapidly transmit them for incorporation into the nuclear DNA before they disintegrate. The first stage involves the energetics of endosome function, which is biologically a complicated process (figure 16.1) (Dinh *et al* 2007). Here the energy stored in membrane deformations can be important. The second stage involves endosomal escape. Often this is driven by an electrostatic phenomenon, dropping the pH increases the charge on the virus and causes it to blow apart. Stage three requires the action of motor proteins similar to

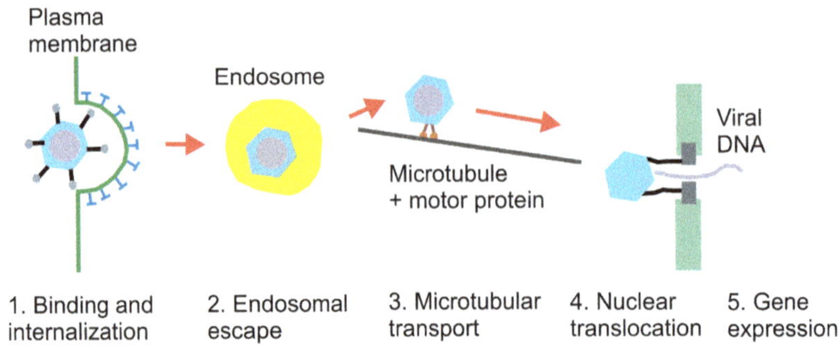

Figure 16.1. Five stages in gene delivery by an adenovirus to the nucleus of a human cell (Dinh *et al* 2007).

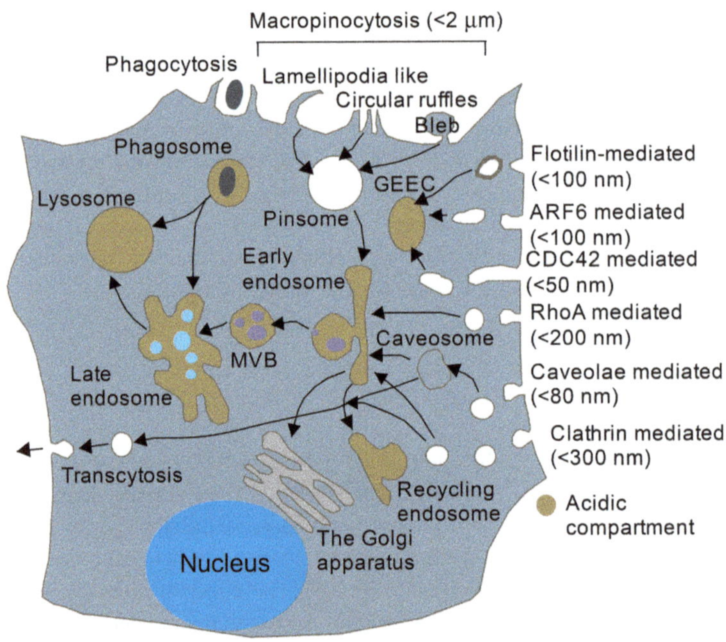

Figure 16.2. Some of the mechanisms used in extracellular uptake during endocytosis in eukaryotic cells (Canton and Battaglia 2012).

those described in chapter two. Stage four involves translocation of a DNA chain through a small pore in the nucleus (again statistical theories, this time for the polymeric conformations of DNA, can be used to understand the process). The final stage involves gene expression, in which a complex range of sub-diffusive motions and reactions are involved (chapters 11 and 12).

16.3.2 Endocytosis

How nanoparticles are imported into cells is crucial in many aspects of modern drug delivery. There are a wide range of mechanisms that naturally occur and some of the principle ones are collected in figure 16.2 (Canton and Battaglia 2012). Viruses are

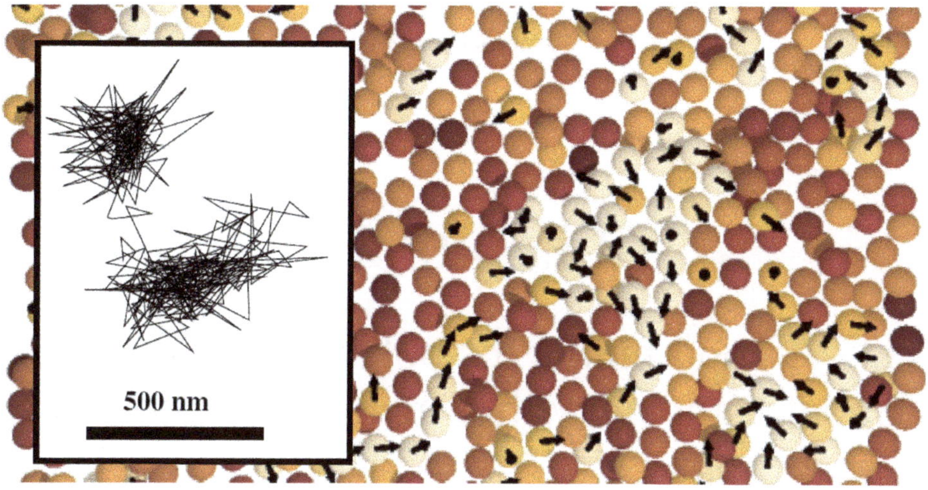

Figure 16.3. A confocal image of a colloidal suspension of PMMA spheres with a radius of 1.18 μm. The more mobile particles are given lighter colours. The inset shows a zoomed in section of a particle track. Similar behaviour is expected with the jamming of colloidal proteins (e.g. globular proteins of 5 nm diameter) which occur at much smaller length scales than the colloids shown. Reprinted with permission from Weeks and Weitz 2002. Copyright 2002 by the American Physical Society.

extremely good at using these import mechanisms, since they need to hijack cells to spread their infections. The majority of particles internalised inside cells are enclosed in membranes and how they do this depends on physical interactions, e.g. the membrane elasticity.

16.3.3 Viscosity of colloidal suspensions and jamming

The variety of phenomena associated with the viscoelasticity of colloidal suspensions (formulations of nanoparticles) is extremely rich (Mewis and Wagner 2011). When traffic jams occur in cities the flow of vehicles can become negligible or oscillate around a reduced value. Similarly, in the flow of jammed colloids at the microscale extremely large viscosities can occur, which results in very low flow rates (figure 16.3). High concentrations of globular proteins (e.g. antibodies) are also susceptible to jamming behaviour, which can be a major barrier to drug delivery.

References

Canton I and Battaglia G 2012 Endocytosis at the nanoscale *Chem. Soc. Rev.* **141** 2178

Dinh A T, Pangarkar C, Theofanous T and Mitragotri S 2007 Understanding intracellular transport processes pertinent to synthetic gene delivery via stochastic simulations and sensitivity analyses *Biophys. J.* **92** 831–46

Mewis J and Wagner N J 2011 *Colloidal Suspension Rheology* (Cambridge: Cambridge University Press)

Weeks E R and Weitz D A 2002 Properties of cage rearrangements observed near the colloidal glass transition *Phys. Rev. Lett.* **89** 957041

Chapter 17

A good model for polyelectrolytes—bootstrapping with a many-body problem

17.1 Introduction

Historically a series of advances have been achieved in the description of the physical world around us, through the introduction of forces laws that describe the motion of objects relative to one another. Famously, a falling apple was supposed to have inspired Isaac Newton to postulate his law of gravity, which allowed him to explain many aspects of planetary motions with the introduction of the concept of forces. Modern biographies of Newton ascribe the story of the falling apple to his niece and claim that Newton himself (a difficult complex individual) denied the story, e.g. read *Isaac Newton* by James Gleick. The falling apple story, although memorable and informative, thus appears to be apocryphal.

Newton's calculations on planetary motion considered only two bodies, say the Earth and the Sun, which only allowed him to make predictions on their orbits with respect to one another. However, what happens when a third body, say the Moon, is included in the calculations? Unfortunately, the calculations become much more of a challenge, and it was eventually proven at the start of the 20th century that no exact solution exists. Therefore, the physical model inferred from Newton's theory, which is really neat and elegant with two bodies, has become much more unwieldly and problematic when additional bodies are added to the system. This is called the *many-body problem.*

People might therefore start to worry about an Armageddon scenario where the Moon crashes into the Earth due to instabilities in the three-body equations of motion (when the Sun, the Earth and the Moon are considered as a closed system). However, scientists (e.g. Henri Poincaré received a prize from King Oscar II in 1887 for his achievements in this area) have been able to prove the stability of the three-body orbits in our solar system (and higher numbers of bodies) through the development of new numerical techniques, and thankfully the possibility of this

doi:10.1088/978-0-7503-1374-2ch17

© IOP Publishing Ltd 2017

particular type of disaster appears to have been averted; at least for now. Similar many-body problems with a long-range interaction raise their ugly heads in numerous other parts of physics, not just those involved with gravity, and there are important examples in many areas of biophysics. Unfortunately, the many-body problem can represent a brick wall that quantitative research hits and then makes little subsequent progress afterwards.

The simplest polymers consist of strings of identical primitive units (monomers) that are glued together with covalent bonds. Biopolymers are often heteropolymers, where a relatively small selection of different primitive units are strung in a line, one after another. Polymers in which the monomers are charged are called *polyelectrolytes* and the repulsion of the charges by their neighbouring stretches of chain in solution gives them unusual physical properties. For example, polyelectrolytes will migrate towards an oppositely charged electrode when placed in an electric field and they migrate at a rate that is inversely proportional to their length (that is in turn proportional to the amount of charge they carry). This has been invaluable for the structural elucidation of biopolymers, using the technique of gel electrophoresis.

The polyelectrolyte class of molecules includes many important biological examples, such as proteins, nucleic acids and carbohydrates. Polyelectrolytes thus provide an example of a many-body problem that represents a huge number of biological systems. The majority of biological molecules are therefore ruled by a many-body problem under the guise of long-range charge interactions. Inclusion of some salt (e.g. sodium chloride) can help matters in polyelectrolyte systems, since it reduces the length scale over which the electrostatic interactions are felt (it screens the charges), but still numerous interactions with neighbouring charged groups need to be accounted for in the calculations and, in general, they appear to be intractable. The salt concentrations are reasonably high in living organisms, e.g. a saline drip is at 0.154 M to balance the salt concentration in blood, but the charged interactions are never fully screened. The majority of biological membranes are also very highly charged and they experience analogous interactions to polyelectrolytes in two dimensions (i.e. in contrast to polyelectrolytes which are one dimensional beasts) and these charged membranes also require quantitative modelling. Specifically, the main components of human membranes are phospholipids that all carry a negative charge, and thus these membranes have an extremely high surface charge density.

The author spent two years of his life working on polyelectrolytes in the de Gennes laboratory of the Collège de France, Paris. Pierre-Gilles de Gennes won the Nobel prize for physics in 1991 for his contributions to the theory of soft condensed matter. For example, his theory of reptation for the dynamics of concentrated solutions of polymers is widely used. The Collège is one of the most prestigious institutions in France; a two-minute walk from the Pantheon (a huge mausoleum packed with the remains of famous French people) and opposite the Sorbonne University. Think of lots of polished marble, Grecian columns and iconoclastic white-haired Nobel Prize winners. Current models for polyelectrolytes were judged to be unsatisfactory, which is why the author was given a two-year contract to work on them as a Maître de Conférence (grandiosely this translates into English as 'Master of the Conferences', but post-doc is a better synonym). The most successful models for polyelectrolytes were (in the late 1990s), and still tend to be, orders of magnitude scaling relationships, and the nagging suspicion is that they are not particularly realistic descriptions of the phenomena involved. Current polyelectrolyte models often provide reasonable numerical estimates of quantities measured in experiments, but without them nailing down the numbers to a few decimal places of accuracy, it is hard to be completely confident that the correct behaviour has indeed been identified. The theories are sometimes off by orders of magnitude when carefully compared with carefully performed experiments.

The polyelectrolyte problem is in part due to the vast variety of interaction strengths possible within the charged molecular systems that occur in water, the main solvent that occurs in nature (humans are 70% water by mass). People often describe charge interactions, not through the use of the number of charges (or equivalently the hydrogen ion concentration), rather they use the logarithm of this number, since practically concentrations can occur over a huge range, $1\text{–}10^{14}$, and it is much easier to confine oneself to numbers from 1–14. This logarithmic scale is used to define the pH of a solution and it is familiar to all professional chemists. 10^{14} is a really big factor, equivalent to the magnification of a bacterium, so it is the size of the distance between the Earth and the Sun. This emphasises the huge range of interaction strengths possible in aqueous solutions due to electrostatics.

The distributions of charges on small length scales are ruled by quantum mechanics. Therefore, the ideal scenario is that models for polyelectrolytes can be coarse grained (averaged) from quantum mechanical foundations. Nevertheless, quantum mechanical analogues of the many-body problem still exist and the consideration of quantum mechanical aspects of the problem does not really help with the fundamental issue of the long-ranged charged interactions (chapter 14). Unfortunately, quantised charged interactions are as hard to handle theoretically as classical charged interactions.

Modern day numerical methods to describe the many-body problem work best with weak forces and small numbers of particles. Other theoretical methods also work well if the particles are perfectly arranged in symmetrical positions in space, e.g. they are organized on lattice sites in a crystal and their behaviour analysed with density functional theory. The regular lattice motif (analogous to that seen in wallpaper) allows electrostatic calculations to be simplified in this case for perfect crystals. However, when a strong long-range interaction is present (such as the electrostatic interactions inside a polyelectrolyte chain), and many disordered particles are considered (all the different charged units along a polyelectrolyte chain), researchers are left with a really hard (in general impossible) problem. Therefore, there is a need to create some smart new ways to average (coarse grain) the behaviour of polyelectrolytes on the mesoscale (the length scales between atoms and the macroscale).

L. *Question.* Design an accurate model of a polyelectrolyte that describes the morphologies adopted by the charged flexible chains.

Once a good model for the static structure of polyelectrolytes has been achieved, the dynamics of the molecules hold a number of important questions, e.g. how DNA chains move in an electrophoresis experiment or how hyaluronan determines the viscosity of synovial joints in people's knees.

L. *Question.* Create an accurate model for polyelectrolyte dynamics that can describe the motion of both long and short polyelectrolyte chains.

The emphasis on short charged chains is there in the question, because they perform particularly badly in current theories.

The introduction of new instrumentation in the field of polyelectrolytes should hopefully encourage some new developments with charged polymers. Even if the many-body problem is in general intractable, experimentally determined solutions (empirical solutions) can sometimes bridge the logical gaps in calculations and allow concrete predictions to be made from theories. This process of phenomenological bootstrapping is in practice a very common one in science. It might make a good honest, but unpronounceable, name for a laboratory, i.e. *The Institute of Phenomenological Bootstrapping*. For example, dielectric spectroscopy is now possible that is based on atomic force microscopes (AFM) and it can provide maps of the electrostatic interactions of molecules at the nanoscale. An AFM consists of a pointed cantilever that is scanned across the surface of a sample using a piezoelectric crystal. The position of the cantilever is measured using light reflected from the back of the cantilever onto a photodiode and at the same time the voltage between the sample and the cantilever is measured. Maps of the dielectric constants of individual molecules and cells can thus be created, and they can then be compared with high resolution simulations for the localized charged interactions experienced by the molecules. This type of AFM experiment could therefore be used to bridge some of the logical gaps in our theoretical understanding of the many-body problem that occurs with polyelectrolytes, e.g. between length scales of 0.1 nanometre and 100 nanometres.

Another subtle issue associated with charge interactions in solution is that of *counterion condensation*. For example, consider a chain of DNA. DNA is a highly charged polyelectrolyte due to the phosphate groups that poke out into the surrounding solution that are attached to the double helical backbone. Three quarters of the phosphate charges along the backbone of DNA chains are neutralized due to counterion condensation. DNA chains thus have significantly reduced charge interactions when they interact with other molecules in solution than might be expected from a casual inspection of their molecular structure. With all highly charged nanoparticles in water, including DNA, there is an extremely high repulsive energy penalty when lots of like-charges are arranged side-by-side and oppositely charged ions that are free in solution can condense onto the particles to reduce this energetic penalty (the free ions lose a little entropy in this process of condensation which is a barrier to its progress, but it is offset by the reduced energy associated with charge repulsion). Counterion condensation is strongly affected by the geometry of charged objects. Different condensation behaviours are observed in one dimension (polymers), two dimensions (membranes) and three dimensions (colloidal particles, such as globular proteins).

L. *Question.* Create an accurate quantitative model for counterion condensation for charged polymers and compare it with some new high-resolution experiments, e.g. dielectric atomic force microscopy (AFM) spectroscopy.

The folding patterns of charged polymers also present some challenges for a quantitative description. Polymers can collapse in on themselves to form a spherical globule if the temperature, pH, pressure or solvent are modified, which is called the *globule-coil transition.* As the chains become globular, huge changes in the electrostatics are experienced as counterions condense onto the globular chains in an avalanche-like process. The physics involved is also observed in nuclear fusion, where a huge electrostatic barrier must be overcome (or nuclear fission as an analogy for the reverse globule-coil process) and it is practically important in the phenomenon of protein folding. This process also occurs with macroscopic liquid droplets when they are charged (e.g. water drops with a static electrical charge), where it is called the Rayleigh charge instability, named after a 19th century English physicist. The instability can cause the droplets to explode into lots of smaller droplets due to electrostatic repulsion, if the charge density exceeds a threshold value.

There is a wide range of other more elaborate phenomena associated with the structure and dynamics of polyelectrolytes that also lack good quantitative theories, such as electrophoresis (the motion of charged polymers in electric fields), zeta potentials (the potential on charged polymers as they move in an electric field), inter-particle potentials (e.g. whether there is an attractive interaction between like-charged polymers), surface tension, viscosity (chapter 13), dielectric responses (how biological materials screen electric fields), drug/repressor/promoter binding to DNA (which is important for gene regulation, chapter 11), gelation, and surface adsorption. Clearly there are a huge range of open questions that relate to the physics of polyelectrolytes.

The attractive interaction between like-charged nanoparticles, such as polyelectrolytes, (presumably an entropic effect) is hotly contested in the literature. Experiments are suggestive that it does indeed exist, but they are hard to reproduce between laboratories and theories lead to few concrete verifiable predictions. Furthermore, fairly ad hoc assumptions are currently made to calculate the surface tension of charged solutions in terms of Lewis acid/base interactions. It is hoped that a more fundamental description is possible. Polyelectrolytes can be switched from liquid-like to solid-like behaviour at very low concentrations, which is important in many technologically relevant applications, such as nappies (diapers) and food gels. Exactly how this process of gelation occurs, combines such a wide range of subtle interactions that it is still a predominantly empirical field. All of these physical questions could benefit from improved models of polyelectrolyte structure, dynamics and counterion condensation.

In comparison with other areas of classical physics (e.g. planetary dynamics, electromagnetism or acoustic vibrations), polyelectrolyte theories are still lousy theories; better than nothing, but with significant room for improvement. Accurate efficient models that allow the calculation of a wide range of charge phenomena in water could help to revolutionise a number of critical subject areas in molecular biophysics, such as the dynamics of DNA chains and phospholipid membranes, but they would also benefit more applied areas of polyelectrolyte research that include gene therapy (chapter 10), osteoarthritis, and heart disease (chapter 18).

17.2 Suggested reading

Introductory

Unfortunately there is no good up-to-date introductory material on polyelectrolytes, in part due to the contentious nature of some of the subject matter. However, there are some undergraduate books on biophysics that have a reasonable introduction, e.g. Philips R (ed) 2013 *Physical Biology of the Cell* (New York: Garland) and Waigh T A 2014 *The Physics of Living Processes: A Mesoscopic Approach* (Chichester: Wiley); or Daune M 1999 *Molecular Biophysics: Structures in Motion* (Oxford: Oxford University Press.)

Gleick J 2004 *Isaac Newton* (New York: Harper Perennial). A well-balanced biography of the famous physicist.

Advanced

Dobrynin A V and Rubinstein M 2005 Theory of polyelectrolytes in solutions and at surfaces *Prog. Polym. Sci.* **30** 1049–118. An advanced theoretical account of modern models for charged polymers.

Dobrynin A V 2015 *The Oxford Handbook of Soft Condensed Matter* ed. E M Terentjev and D A Weitz (Oxford: Oxford University Press). Some more progress with the problems involved in polyelectrolyte physics is described

Esteban-Ferrer D, Edwards M A, Fumagalli L, Juarez A and Gomila G 2014 Electric polarization properties of single bacteria measured with electrostatic force microscopy *ACS Nano* **8** 9843–9.

Cuervo A, Dans P D, Carrascosa J L, Orozco M, Gomila G and Fumagalli L 2014 Direct measurement of the dielectric polarization properties of DNA *PNAS* **111** E3624–E30. These primary research articles describe the new technique of dielectric spectroscopy based on atomic force microscopes (AFMs), which can provide spatially resolved images of the dielectric constants of both single molecules and bacteria.

17.3 Technical details

17.3.1 Scaling model of polyelectrolytes

Scaling models for polyelectrolytes find their origin in the work of the Nobel prize winning physicist, Pierre Giles de Gennes (de Gennes 1979). He used the blob concept in a range of scenarios to provide coarse-grained models for the conformations of polymers. In polyelectrolytes, *electrostatic blobs* are defined in which the electrostatic interactions experienced by a section of the chain have an energy equal to the thermal energy. The conformation of a polyelectrolyte is thus a rod of charged blobs (figure 17.1).

When the rod-like assemblies of blobs overlap, charge screening allows them to deviate from rod-like conformations at length scales larger than the mesh size of the network. More details on polyelectrolyte blob models can be found in Dobrynin *et al* (Dobrynin and Rubinstein 2005). These blob models capture some of the

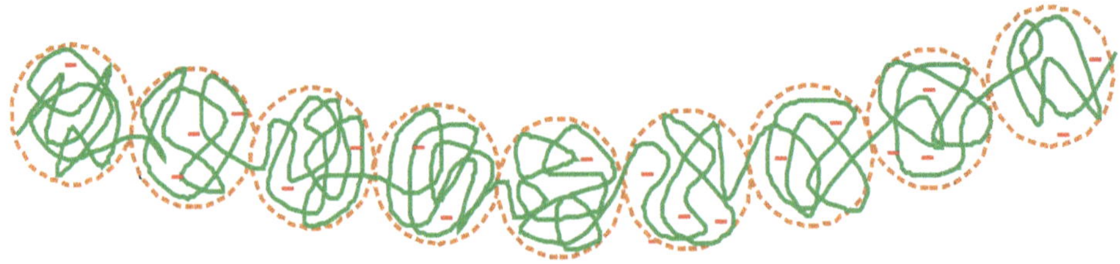

Figure 17.1. Schematic diagram of the conformation of a flexible charged polymer (a rod of orange blobs), e.g. DNA. The chain is deformed slightly at long length scales implying some additional energy has been stored in it.

crucial features of the physics of polyelectrolytes, but they are not fully quantitative and fail in some situations, e.g. for short chains.

17.3.2 Electrophoresis

A number of important technological phenomena depend on the dynamics of polyelectrolyte chains (e.g. polyelectrolyte rheology, chapter 13). Here, electrophoresis is highlighted, since it plays a central role in genetics for DNA sequencing and thus modern molecular biology (it therefore plays a key role in gene finger printing, mutagenesis etc).

The experimental requirements for the simplest electrophoresis experiment are relatively straightforward. They require a polyacrylamide gel, two electrodes, a polyelectrolyte sample, a buffer, a voltage supply and a method to detect the position of the polyelectrolyte as it moves across the gel, e.g. fluorescence labels. The DNA (or other polyelectrolyte) is pulled through the gel as it migrates towards the electrode of opposite polarity. The rate at which the DNA travels primarily depends on its length (the shape effects are often negligible, since the chains are linearized before experiments), but also on subtle hydrodynamic interactions with the gel. The measurement of length from the transit times makes possible the sequencing of the polyelectrolytes when combined with additional biochemical reactions, e.g. termination reactions in shot gun sequencing (Waigh 2014).

17.3.3 Rayleigh charge instability

An interesting challenge for polyelectrolyte theories is found in materials sensitive to the Rayleigh charge instability (Dobrynin *et al* 1996). The effect is due to the tendency of polymers that dislike the solvent in which they are dissolved (e.g. hydrophobic polymers) to adopt globular structures. If the chains are also charged, there is a driver that causes the opposite effect, the chains tend to stretch out in the solution due to charge repulsion. In between the extremes of the single globule and the fully stretched conformations, intermediate morphologies are observed with multiple globules arranged on a single chain (figure 17.2). Similar phenomena occur in the break-up of fluid jets (the system Rayleigh originally considered) and nuclear fission.

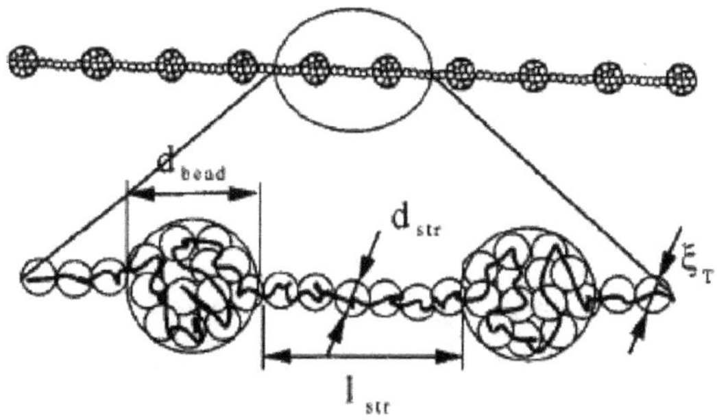

Figure 17.2. Rayleigh charge instability in hydrophobic polyelectrolytes can lead to a series of globules attached in a line (pearls on a necklace morphology). Reprinted with permission from Dobrynin *et al* 1996. Copyright 1996 American Chemical Society.

References

de Gennes P G 1979 *Scaling Concepts in Polymer Physics* (Ithaca, NY: Cornell University Press)

Dobrynin A V and Rubinstein M 2005 Theory of polyelectrolytes in solutions and at surfaces *Prog. Polym. Sci.* **30** 1049–118

Dobrynin A V, Rubinstein M and Obukhov S O 1996 Cascade of transitions in polyelectrolyte chain in poor solvents *Macromolecules* **29** 2974–9

Waigh T A 2014 *The Physics of Living Processes: A Mesoscopic Approach* (Chichester: Wiley)

IOP Publishing

Some Critical Questions in Biological Physics
A guided tour around the bugbears
Thomas Andrew Waigh

Chapter 18

The activity of hearts—the pump that quivers

18.1 Introduction

The author's research into the biophysics of hearts has been relatively limited, more due to historical quirks in funding rather than any lack of any real interest in the problems involved. A number of years ago some simple projects were completed on the structure, dynamics and elasticity of biomolecules from heart tissue (titin and thin filaments). Titin is the largest single protein that occurs in human bodies and provides passive tension to pull the lattice of motor proteins in striated muscle (e.g. in cardiac muscle) back together after a contraction. Thin filaments form railroad tracks along which the nanomotors in heart tissue (myosins) can run. However, more recently there has been lots of contact with a successful colleague here in Manchester's School of Physics and Astronomy, who has a large cohort of students that work on the computational modelling of the electrophysiology of hearts. Students thus have lots of exposure to questions in cardiac research in our biological physics group.

The sobering statistic is that heart disease is the number one killer in the world. As such, it is the predominant mechanism that will cause the reader and all their friends, relatives, neighbours and associates to eventually pop their clogs. This is a slightly grim statistic and a quick visit to a GP's surgery or a local hospital will confirm its truth. Doctors' surgeries are packed full of individuals that have a panoply of heart associated diseases. For example, babies born with holes in their hearts, ostensibly healthy athletes with dangerously erratic heart beats or heart attack patients that have stents inserted into them to give a few more precious years. The motivation behind cardiac research therefore does not need much salesmanship, although it tends to be fairly conservative as a rule. Flamboyant heart surgeons who are overly creative are unsurprisingly not in great demand. Research historically has been desperation led, with heavy burdens of responsibility on the researchers to come up with quick fixes to fatal medical conditions, within some fairly limited time

doi:10.1088/978-0-7503-1374-2ch18 © IOP Publishing Ltd 2017

constraints. A crude mechanical engineering solution to heart disease just involves a peristaltic pump and a piece of polyurethane tubing, but not much thought is required before clear problems are detected with this naïve solution.

Hearts are electromechanical devices in which both electrical and mechanical mechanisms together play a key role in the organization of the contractions that pump blood around the body. Human hearts consist of two pumps that are stuck together as one unit. Other organisms can have different numbers of hearts, e.g. squid have three separated pumps, i.e. three hearts. *In utero* human babies only use one of the pumps in their hearts (they do not need to pump blood around their lungs to absorb oxygen in the womb) and the second pump only begins to circulate blood upon birth (a hole in the heart closes during birth). Electrical effects drive the timing of the contraction of the pumps (poorly organized timing can lead to limited blood circulation, no blood circulation or, even worse, circulation in the wrong direction), whereas the mechanical components are needed for the actual action of the pumps. Breaking the problem down into its component parts provides a first target for an improved understanding; the electrical effects need to be understood on the level of a single cardiac cell.

L. *Question:* Create an accurate spatially resolved model for intracellular conductivity in a single cardiac cell.

The creation of accurate electrophysiological models of the space inside heterogeneous cardiac cells is challenging. Ion conductivity has marked changes from point to point in cells, which often have lots of internal compartments. Furthermore, there are still lots of gaps in the understanding of the fundamental biology of cells in general, and thus also in cardiac cells, e.g. how the sarcoplasmic reticulum works on the nanoscale (the equivalent of the endoplasmic reticulum for cardiac cells, chapter 15). The process of electrophysiological modelling will be helped by the development of new imaging techniques, such as super-resolution fluorescence microscopy or electrostatic force microscopy. These will enable changes in electrical activity to be correlated with the underlying biochemical circuits. The effect of feedback from the mechanical behaviour of the muscular contractions on the electrical cardiac waves also needs to be considered.

Once a good model exists for a single cardiac cell, the obvious next step is to create a model for the electrical waves that travel across the entire organ.

L. *Question:* Combine all the different types of cardiac cell into an anatomically accurate model of the electrical activity observed across a whole heart.

In silico models provided by computer based simulations ease up on ethical dilemmas with animal research and they can improve our understanding of how electrical waves are conducted around whole hearts. Nevertheless, care is needed to insure results are both medically relevant and well based in reality, or lots of funding will be wasted on silicon based fantasies.

One interesting new perspective for quantitative models is the discovery of the sub-diffusive motion of electrical waves as they move around the heart. To date, the majority of theories assume purely diffusive modes of transport (using reaction-diffusion equations), but experiments give clear evidence for slightly slower sub-diffusive transport. This unusual form of sub-diffusive motility requires fractional differential equations for its description (chapter 12) and could provide a breakthrough in the quantitative understanding of electrical waves in heterogeneous heart tissue.

From the perspective of fundamental physics, Maxwell's equations, the fundamental laws that describe all electromagnetic phenomena in Nature, must also determine the electrical phenomena associated with cardiac waves. Heart research thus appears to be just a problem of modelling all the different electrical mechanisms that are involved using Maxwell's equations. The cardiac simulation effort therefore is currently mostly targeted at ion channels that conduct currents through the membranes of heart cells. Models use simple electrical circuit theory to describe the currents through individual ion channels and then use Maxwell's equations to predict how these currents relate to the electrical waves measured across whole hearts, e.g. using electrocardiograms (ECGs). ECGs are a standard method to monitor people's vital signs in a hospital. Standard high-resolution ECG measurements often involve 12 electrodes (the minimum is three) and the electrodes are placed all over a subject's body. Excitable cells in the heart provide the largest contribution to the body's electrical potentials and they are conducted fairly well over the entire surface of the human body. The ECGs can thus be interpreted in terms of waves emitted by the heart's pacemaker (the sinoatrial node) that then propagate across the rest of the organ. Diseased states can be identified if the time series of the ECG does not propagate in the correct order or with the correct periodicity, so that efficient contractions of the two pumps in the heart are not achieved.

Non-linear physics is a very subtle and intricate field, which includes sophisticated phenomena such as chaos, turbulence (chapter 5) and fractals (chapter 12). Many biological processes are non-linear and the non-linearities in these systems are often finely optimised for their tasks. In fact, it can be a challenge to find many biological processes that are truly linear, since most mechanical properties, reaction kinetics (predominantly enzyme driven), fluid mechanics (the flow of blood) and electrical activity (the spikes in a neuron) that occur in biology are fundamentally non-linear. Lots of non-linear phenomena occur within the electrical patterns transmitted through cardiac tissue, which provides the organs with a rich variety of dynamical states, e.g. chaotic patterns are associated with some diseases that cause heart failure, such as ventricular fibrillation. The ventricules are the lower sections of the two pumps in human hearts (the left and right sections) and fibrillation is the fast irregular contractions of muscle fibres (a quivering motion) that occur in heart attacks.

Picture a Hollywood operation room where the actors are all dressed in doctors' laboratory coats and the lead actor shouts 'clear' before zapping the patient with thousands of volts using some electrical paddles. This is the procedure of *heart defibrillation* which is used in hospitals worldwide to restart people's failing hearts. An immediate application for cardiac modelling is therefore seen in defibrillation. The high voltages used to restart hearts can cause significant amounts of damage to people's bodies as well as saving their lives. Clearly there is work required to reduce the voltages to safer levels, whilst still restarting the electrical activity to allow the cardiac muscles to contract in the correct order. Recent modelling work has provided new insights to reduce the amount of electrical energy that is required to control electrical turbulence in the heart and it has thus provided inspiration for the design of new defibrillators.

Moving on to mechanical aspects of cardiac activity, it is intriguing to consider whether the *viscoelasticity* of cardiac tissue varies across the whole organ. In general, viscoelasticity describes how materials flow and respond to deformation. Viscoelastic flows depend on the time scale at which they are explored. A classic example of viscoelasticity is silly putty that acts as an elastic solid at short times (it bounces when thrown on the floor) and a fluid at long times (when pulled slowly in the hand). Hearts also display viscoelastic phenomena when they contract, although they tend to be much more elastic than silly putty. In principle, modern microrheology methods (chapter 13) can allow this viscoelasticity to be probed in a heart with better resolution than ever before, e.g. at different positions within a single cardiac cell.

L. *Question:* Measure the viscoelasticity of a single cardiac cell over a wide range of frequencies at different positions inside the cell.

This viscoelastic data could then be used to inform better mechanical models of the heart and how these relate to diseased states, e.g. when localized regions of the organ die and become more rigid which is called a cardiac infarction (chapter 13). Heart muscle is unusual in that it can demonstrate *active viscoelasticity*. The activity of the motor proteins inside cardiac muscle actively modulates the viscoelasticity of the organ (they tend to soften it) at the same time as they affect the pumping efficiency, the available forces, the blood pressure etc.

Cardiac research is a wide-ranging speciality in medicine, but research is often mainly descriptive and polarised around the traditional areas of the pharmacology of ion channels and the overlapping fields associated with new developments in the molecular biology of cardiac cells. However, some quantitative research has been performed, which hints that a more fundamental understanding is possible, and it could be used to form quantitative tools to combat disease. The ideal scenario would be to bridge all the length scales from single molecules (e.g. ion channels and motor proteins) to whole heart physiology describing both its electrical and mechanical activity simultaneously, i.e. an integrative model of the heart.

The insertion of a stent into a heart is now a very standard procedure to treat heart disease in hospitals and the operation is relatively uncomplicated. If you believe heart surgeons, they just line up the heart attack patients on gurneys every morning and steadily work through the operations for each individual during the day. An incision is made in someone's thigh and then the stent is poked up an artery into the blocked region of the heart's muscle. The blood supply is re-established to the heart muscle when the stent is expanded, and the heart can then restart an efficient process of pumping.

Stents are very successful treatments, but issues with blockages can occur later in people's lives as material from the blood adheres to the foreign surface of the stent and this again will compromise the activity of the heart. There is therefore a need to make a hard-wearing coating for a heart stent that resists both blood clotting and other sources of blockage, such as the attachment of bacterial biofilms (chapter 10), which is the second most important mechanism of stent biofouling. Graphene is a new wonder material that led to the award of the Nobel Prize in physics to Andre Geim and Kosta Novslov in 2013 at the University of Manchester. One current possibility is to paint the surface of stents with graphene oxide, since graphene based materials have interesting hard wearing mechanical properties. Whether graphene films will have useful medical applications is yet to be proven.

L. *Question:* Create a non-stick coating for a stent that resists the adhesion of both blood clots and bacterial biofilms.

Blood clots are a bit like biomolecular Araldite (the two reacting components are thrombin and fibrinogen) that sticks to most molecular surfaces, so it is no easy task to stop the processes of adhesion that cause stent biofouling. Bacteria are also very efficient at gluing themselves onto surfaces, although a variety of molecular mechanisms can be used to prevent this process of adhesion. The modulation of the forces experienced by the surface of the stents will thus play an important role in the creation of effective non-stick coatings.

Clearly there are numerous important applications of biophysics in cardiac medicine. Effective treatments for heart disease (stents, blood thinners, beta blockers etc) have caused a major improvement in the mortality rates experienced by people around the world. Cardiac research is perhaps unusual in biological physics in that it is such a well-defined problem. In principle, we just need to fit a good biocompatible peristaltic pump as a replacement for failing hearts or treat the ones we are born with slightly more carefully, so they last longer. In the second case it is assumed that the person has been lucky enough to be supplied with a functional heart at the time of their birth. Making the replacement pumps and all of their internal components completely biocompatible is perhaps the main issue that prevents progress in heart research.

It is hoped that advances in cardiac research will continue to improve people's quality of life and that some of these advances will be driven by quantitative biological physics research, e.g. new heart defibrillators, new stent architectures or new drugs to modulate the activity of ion channels will become possible.

18.2 Suggested reading

Introductory

Noble D 2008 *The Music of Life: Biology Beyond Genes* (Oxford: Oxford University Press.) The author was one of the first people to simulate the electrical activity of the heart on a computer and this type of research is placed in the wider context of systems biology in this book.

Feher J 2012 *Quantitative Human Physiology: An Introduction* (Cambridge: Academic). A general introduction to heart physiology.

Advanced

Keener J and Sneyd J 2009 *Mathematical Physiology* (Berlin: Springer). This book tends to be the standard text for researchers who build models of hearts based on the Hodgkin–Huxley equations to describe electrophysiological phenomena.

Luther S *et al* 2011 Low-energy control of electrical turbulence in the heart *Nature* **475** 235. Primary research literature that describes new strategies to control heart activity using low voltage defibrillation.

Coleman M A, Aslanidi O V, Kharche S, Boyett M R, Garratt C, Hancox J C and Zhang H 2013 Pro-arrhythmogenic effects of atrial fibrillation-induced electrical remodelling: insights from the three-dimensional virtual human atria *J. Physiol* **591** 4249–72. Primary research literature that describes state-of-the-art simulations of the electrical activity across a heart.

Strogatz S H 2014 *Non-Linear Dynamics and Chaos* (Boulder, CO: Westview). Classic introduction to the mathematics of non-linear phenomena.

18.3 Technical details

18.3.1 Anatomy of a heart

Figure 18.1 shows a simple schematic diagram of the anatomy of a heart (Feher 2012). As shown, the left (blue) side pumps blood around the lungs and the right (red) side then circulates this oxygenated blood around the rest of the body. An analogous picture can be drawn showing the electrical components of the heart, and they have a well-defined order of excitation, as electrical waves guide the consecutive contractions of different mechanical components.

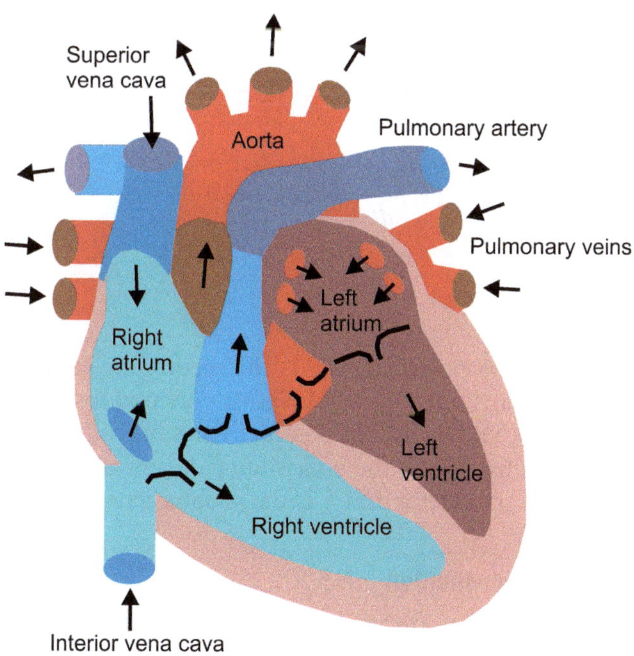

Figure 18.1. The simple anatomy of the human heart that shows the principle mechanical components.

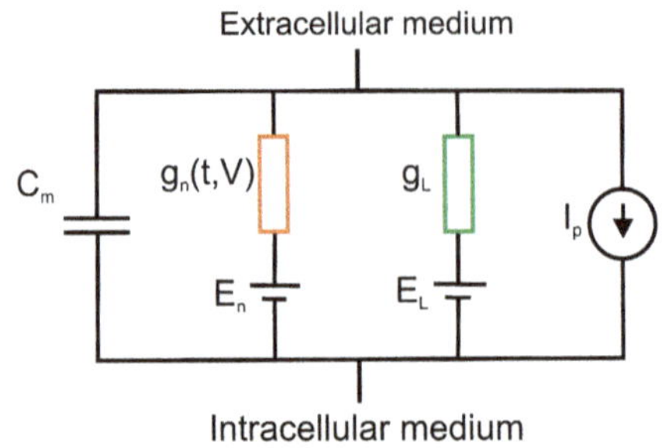

Figure 18.2. A simple electronic circuit model for the membranes of excitable cells that form the basis of the Hogkin–Huxley model. $g_n(t, V)$ is the conductance of ion channel n (either sodium or potassium) as a function of time (t) and voltage (V). E_n and E_L are the EMFs of the ion channel and the leakage channel EMFs respectively. I_p is the applied current. C_m is the membrane capacitance.

18.3.2 Hodgkin–Huxley equations

The fundamental model for the electrical activity of neurons was provided by Hodgkin and Huxley (Hodgkin and Huxley 1952). Four equations were proposed in the original model and the equivalent electrical circuit is shown in figure 18.2. The first equation describes the conservation of current in the equivalent electronic circuit (figure 18.2) and the additional three equations describe the time responses of the sodium and potassium ion channels.

$$C_m \frac{dV}{dt} + g_K n^4 (V - V_K) + g_{Na} m^3 h (V - V_{Na}) + g_L (V - V_L) + I_{app} = 0 \quad (18.1)$$

$$\tau_n(V) \frac{dn}{dt} = n_\infty(V) - n \quad (18.2)$$

$$\tau_h(V) \frac{dh}{dt} = h_\infty(V) - h \quad (18.3)$$

$$\tau_m(V) \frac{dm}{dt} = m_\infty(V) - m \quad (18.4)$$

where V is the membrane potential, C_m is the membrane capacitance, V_K is the Nernst potential for potassium, V_{Na} is the Nernst potential for sodium and V_L is the leakage voltage. g_K, g_{Na} and g_L are the conductivities of the potassium, sodium and leakage channels respectively. n is the potassium gate activation parameter, m is the sodium gate activation parameter and h is the sodium gate inactivation parameter.

18.3.3 Reaction–diffusion equation

Reaction–diffusion equations are used to describe the motion of electrical waves across the surface of hearts. The waves are often assumed to be described by the mono domain model and, as such, the dynamics are assumed to be driven by simple diffusion, which is known to be unsatisfactory in many forms of intracellular transport (chapter 12). Specifically, the monodomain model is given by

$$\nabla.\,(\sigma\nabla V_{\mathrm{m}}) = \chi(C_{\mathrm{m}}\partial_{\mathrm{t}}V_{\mathrm{m}} + I_{\mathrm{ion}} - I_{\mathrm{stim}}) \tag{18.5}$$

where V_{m} is the cellular trans membrane potential, σ is a conductivity tensor, χ is the cell surface-to-volume ratio and C_{m} is again the membrane capacitance. I_{ion} and I_{stim} are the total transmembrane and stimulus currents, respectively.

Can the assumption of simple diffusion be improved upon? For isotropic conduction a fractional monodomain equation is

$$\partial_{\mathrm{t}}V_{\mathrm{e}} = -D_{\alpha}(-\Delta)^{\alpha/2}V_{\mathrm{m}} - \frac{1}{C_{\mathrm{m}}}(I_{\mathrm{ion}} - I_{\mathrm{stim}}) \tag{18.6}$$

where D_{α} is the diffusion coefficient and $(-\Delta)^{\alpha/2}$ is a fractional Laplacian operator. Reference should be made to the original literature for numerical solutions to this equation combined with fits to experimental data (Bueno-Orovio *et al* 2014).

References

Bueno-Orovio A, Kay D, Grau V, Rodriguez B and Burrage K 2014 Fractional diffusion models of cardiac electrical propagation: role of structural heterogeneity in dispersion of repolarization *J. R. Soc. Interface* **11** 1–12

Feher J 2012 *Quantitative Human Physiology: An Introduction* (Cambridge: Academic)

Hodgkin A L and Huxley A F 1952 A quantitative description of membrane current and its application to conduction and excitation in nerve *J. Physiol.* **117** 500–44

IOP Publishing

Some Critical Questions in Biological Physics
A guided tour around the bugbears
Thomas Andrew Waigh

Chapter 19

A perspective

> 'We're all born bald, baby.'
> Kojak (Telly Savalas)
> An amusing, but technically incorrect statement that could have been easily rectified by additional biophysical research.

In conclusion, it is hoped that a good case has been made that biological physics research is an exciting endeavour. It represents some of the most interesting and important unsolved problems in both physics research and in science in general. Furthermore, there is plenty of work to be done in biological physics for the foreseeable future. Many well-respected scientists have predicted that further earth-shattering discoveries in biology will occur during the 21st century. These discoveries will revolutionise everyone's lives, in a similar manner that physics changed the world in the 20th century with the invention of transistors, satellite communication, magnetic memories, lasers, space craft etc. It is hoped that physics can continue to contribute to this new scientific revolution, and a major part of its influence will be felt through interdisciplinary areas, such as biological physics.

doi:10.1088/978-0-7503-1374-2ch19 © IOP Publishing Ltd 2017

Lightning Source UK Ltd.
Milton Keynes UK
UKHW052321151218
333888UK00004B/49/P

9 780750 313759